COMMON CORE MATHEMATICS

NEW YORK EDITION

Grade 5, Module 1: Place Value and Decimal Fractions

COMMON CORE

JOSSEY-BASS
A Wiley Imprint
www.josseybass.com

Cover design by Chris Clary

Copyright © 2013 by Common Core, Inc. All rights reserved.

Published by Jossey-Bass
A Wiley Brand
One Montgomery Street, Suite 1200, San Francisco, CA 94104-4594—www.josseybass.com

No part of this publication may be reproduced, stored in a retrieval system, or transmitted in any form or by any means, electronic, mechanical, photocopying, recording, scanning, or otherwise, except as permitted under Section 107 or 108 of the 1976 United States Copyright Act, without either the prior written permission of the publisher, or authorization through payment of the appropriate per-copy fee to the Copyright Clearance Center, Inc., 222 Rosewood Drive, Danvers, MA 01923, 978-750-8400, fax 978-646-8600, or on the Web at www.copyright.com. Requests to the publisher for permission should be addressed to the Permissions Department, John Wiley & Sons, Inc., 111 River Street, Hoboken, NJ 07030, 201-748-6011, fax 201-748-6008, or online at www.wiley.com/go/permissions.

Limit of Liability/Disclaimer of Warranty: While the publisher and author have used their best efforts in preparing this book, they make no representations or warranties with respect to the accuracy or completeness of the contents of this book and specifically disclaim any implied warranties of merchantability or fitness for a particular purpose. No warranty may be created or extended by sales representatives or written sales materials. The advice and strategies contained herein may not be suitable for your situation. You should consult with a professional where appropriate. Neither the publisher nor author shall be liable for any loss of profit or any other commercial damages, including but not limited to special, incidental, consequential, or other damages. Readers should be aware that Internet Web sites offered as citations and/or sources for further information may have changed or disappeared between the time this was written and when it is read.

ISBN 978-1-118-79297-1

Printed in the United States of America
FIRST EDITION
PB Printing   10 9 8 7 6 5 4 3 2 1

WELCOME

Dear Teacher,

Thank you for your interest in Common Core's curriculum in mathematics. Common Core is a non-profit organization based in Washington, DC dedicated to helping K-12 public schoolteachers use the power of high-quality content to improve instruction.[1] We are led by a board of master teachers, scholars, and current and former school, district, and state education leaders. Common Core has responded to the Common Core State Standards' (CCSS) call for "content-rich curriculum"[2] by creating new, CCSS-based curriculum materials in mathematics, English Language Arts, history, and (soon) the arts. All of our materials are written by teachers who are among the nation's foremost experts on the new standards.

In 2012 Common Core won three contracts from the New York State Education Department to create a PreKindergarten–12th grade mathematics curriculum for the teachers of that state, and to conduct associated professional development. The book you hold contains a portion of that work. In order to respond to demand in New York and elsewhere, modules of the curriculum will continue to be published, on a rolling basis, as they are completed. This curriculum is based on New York's version of the CCSS (the CCLS, or Common Core Learning Standards). Common Core will be releasing an enhanced version of the curriculum this summer on our website, commoncore.org. That version also will be published by Jossey-Bass, a Wiley imprint.

Common Core's curriculum materials are not merely aligned to the new standards, they take the CCSS as their very foundation. Our work in math takes its shape from the expectations embedded in the new standards—including the instructional shifts and mathematical progressions, and the new expectations for student fluency, deep conceptual understanding, and application to real-life context. Similarly, our ELA and history curricula are deeply informed by the CCSS's new emphasis on close reading, increased use of informational text, and evidence-based writing.

Our curriculum is distinguished not only by its adherence to the CCSS. The math curriculum is based on a theory of teaching math that is proven to work. That theory posits that mathematical knowledge is most coherently and

---

1. Despite the coincidence of name, Common Core and the Common Core State Standards are not affiliated. Common Core was established in 2007, prior to the start of the Common Core State Standards Initiative, which was led by the National Governors Association and the Council for Chief State School Officers.
2. *Common Core State Standards for English Language Arts & Literacy in History/Social Studies, Science, and Technical Subjects* (Washington, DC: Common Core State Standards Initiative), 6.

effectively conveyed when it is taught in a sequence that follows the "story" of mathematics itself. This is why we call the elementary portion of this curriculum "The Story of Units," to be followed by "The Story of Ratios" in middle school, and "The Story of Functions" in high school. Mathematical concepts flow logically, from one to the next, in this curriculum. The sequencing has been joined with methods of instruction that have been proven to work, in this nation and abroad. These methods drive student understanding beyond process, to deep mastery of mathematical concepts. The goal of the curriculum is to produce students who are not merely literate, but fluent, in mathematics.

It is important to note that, as extensive as these curriculum materials are, they are not meant to be prescriptive. Rather, they are intended to provide a basis for teachers to hone their own craft through study, collaboration, training, and the application of their own expertise as professionals. At Common Core we believe deeply in the ability of teachers and in their central and irreplaceable role in shaping the classroom experience. We strive only to support and facilitate their important work.

The teachers and scholars who wrote these materials are listed beginning on the next page. Their deep knowledge of mathematics, of the CCSS, and of what works in classrooms defined this work in every respect. I would like to thank Louisiana State University professor of mathematics Scott Baldridge for the intellectual leadership he provides to this project. Teacher, trainer, and writer Robin Ramos is the most inspired math educator I've ever encountered. It is Robin and Scott's aspirations for what mathematics education in America *should* look like that is spelled out in these pages.

Finally, this work owes a debt to project director Nell McAnelly that is so deep I'm confident it never can be repaid. Nell, who leads LSU's Gordon A. Cain Center for STEM Literacy, oversees all aspects of our work for NYSED. She has spent days, nights, weekends, and many cancelled vacations toiling in her efforts to make it possible for this talented group of teacher-writers to produce their best work against impossible deadlines. I'm confident that in the years to come Scott, Robin, and Nell will be among those who will deserve to be credited with putting math instruction in our nation back on track.

Thank you for taking an interest in our work. Please join us at www.commoncore.org.

Lynne Munson
President and Executive Director
Common Core
Washington, DC
June 20, 2013

Common Core's Trustees

Erik Berg, elementary public school teacher in Boston, Massachusetts
Barbara Byrd-Bennett, Chief Executive Officer of the Chicago Public Schools
Antonia Cortese, former Secretary-Treasurer of the American Federation of Teachers
Pascal Forgione, Jr., Executive Director of the Center on K-12 Assessment and Performance Management at ETS
Lorraine Griffith, elementary public school teacher in Asheville, North Carolina
Jason Griffiths, Principal of the Harlem Village Academy High School
Bill Honig, President of the Consortium on Reading Excellence
Carol Jago, Director of the California Reading and Literature Project at UCLA
Richard Kessler, Dean of Mannes College, The New School for Music
Lynne Munson, President and Executive Director of Common Core
Juan Rangel, Chief Executive Officer of Chicago-based United Neighborhood Organization

Common Core's Washington, D.C., Staff

Lynne Munson, President and Executive Director
Barbara Davidson, Deputy Director
Sandra Elliott, Director of Professional Development
Sarah Woodard, Programs Manager
Rachel Rooney, Programs Manager
Alyson Burgess, Partnerships Manager
Becca Wammack, Development Manager
Lauren Shaw, Programs Assistant
Diego Quiroga, Membership Coordinator
Elisabeth Mox, Executive Assistant to the President

Common Core's K-5 Math Staff

Scott Baldridge, Lead Mathematician and Writer
Robin Ramos, Lead Writer, PreKindergarten-5
Jill Diniz, Lead Writer, 6-12
Ben McCarty, Mathematician

Nell McAnelly, Project Director
Tiah Alphonso, Associate Director
Jennifer Loftin, Associate Director
Catriona Anderson, Curriculum Manager, PreKindergarten-5

Sherri Adler, PreKindergarten
Debbie Andorka-Aceves, PreKindergarten

Kate McGill Austin, Kindergarten
Nancy Diorio, Kindergarten
Lacy Endo-Peery, Kindergarten
Melanie Gutierrez, Kindergarten
Nuhad Jamal, Kindergarten
Cecilia Rudzitis, Kindergarten
Shelly Snow, Kindergarten

Beth Barnes, First Grade
Lily Cavanaugh, First Grade
Ana Estela, First Grade
Kelley Isinger, First Grade
Kelly Spinks, First Grade
Marianne Strayton, First Grade
Hae Jung Yang, First Grade

Wendy Keehfus-Jones, Second Grade
Susan Midlarsky, Second Grade
Jenny Petrosino, Second Grade
Colleen Sheeron, Second Grade
Nancy Sommer, Second Grade
Lisa Watts-Lawton, Second Grade
MaryJo Wieland, Second Grade
Jessa Woods, Second Grade

Eric Angel, Third Grade
Greg Gorman, Third Grade
Susan Lee, Third Grade
Cristina Metcalf, Third Grade
Ann Rose Santoro, Third Grade
Kevin Tougher, Third Grade
Victoria Peacock, Third Grade
Saffron VanGalder, Third Grade

Katrina Abdussalaam, Fourth Grade
Kelly Alsup, Fourth Grade
Patti Dieck, Fourth Grade
Mary Jones, Fourth Grade
Soojin Lu, Fourth Grade
Tricia Salerno, Fourth Grade
Gail Smith, Fourth Grade
Eric Welch, Fourth Grade
Sam Wertheim, Fourth Grade
Erin Wheeler, Fourth Grade

Leslie Arceneaux, Fifth Grade
Adam Baker, Fifth Grade
Janice Fan, Fifth Grade
Peggy Golden, Fifth Grade
Halle Kananak, Fifth Grade
Shauntina Kerrison, Fifth Grade
Pat Mohr, Fifth Grade
Chris Sarlo, Fifth Grade

Additional Writers

Bill Davidson, Fluency Specialist
Robin Hecht, UDL Specialist
Simon Pfeil, Mathematician

Document Management Team

Tam Le, Document Manager
Jennifer Merchan, Copy Editor

GRADE 5 • MODULE 1

Table of Contents
GRADE 5 • MODULE 1
Place Value and Decimal Fractions

Module Overview i

Topic A: Multiplicative Patterns on the Place Value Chart 1.A.1

Topic B: Decimal Fractions and Place Value Patterns 1.B.1

Topic C: Place Value and Rounding Decimal Fractions 1.C.1

Topic D: Adding and Subtracting Decimals 1.D.1

Topic E: Multiplying Decimals 1.E.1

Topic F: Dividing Decimals 1.F.1

Module Assessments 1.S.1

COMMON CORE MATHEMATICS CURRICULUM • NY   Module Overview **5•1**

Grade 5 • Module 1
Place Value and Decimal Fractions

OVERVIEW

In Module 1, students' understanding of the patterns in the base ten system are extended from Grade 4's work with place value of multi-digit whole numbers and decimals to hundredths to the thousandths place. In Grade 5, students deepen their knowledge through a more generalized understanding of the relationships between and among adjacent places on the place value chart, e.g., 1 tenth times any digit on the place value chart moves it one place value to the right (**5.NBT.1**). Toward the module's end students apply these new understandings as they reason about and perform decimal operations through the hundredths place.

Topic A opens the module with a conceptual exploration of the multiplicative patterns of the base ten system using place value disks and a place value chart. Students notice that multiplying by 1000 is the same as multiplying by 10 x 10 x 10. Since each factor of 10 shifts the digits one place to the left, multiplying by 10 x 10 x 10—which can be recorded in exponential form as 10^3 (**5.NBT.2**)—shifts the position of the digits to the left 3 places, thus changing the digits' relationships to the decimal point (**5.NBT.2**). Application of these place value understandings to problem solving with metric conversions completes Topic A (**5.MD.1**).

Topic B moves into the naming of decimal fraction numbers in expanded, unit (e.g., 4.23 = 4 ones 2 tenths 3 hundredths), and word forms and concludes with using like units to compare decimal fractions. Now in Grade 5, students use exponents and the unit fraction to represent expanded form, e.g., $2 \times 10^2 + 3 \times (1/10) + 4 \times (1/100) = 200.34$ (**5.NBT.3**). Further, students reason about differences in the values of like place value units and expressing those comparisons with symbols (>, <, and =). Students generalize their knowledge of rounding whole numbers to round decimal numbers in Topic C initially using a vertical number line to interpret the result as an approximation, eventually moving away from the visual model (**5.NBT.4**).

In the latter topics of Module 1, students use the relationships of adjacent units and generalize whole number algorithms to decimal fraction operations (**5.NBT.7**). Topic D uses unit form to connect general methods for addition and subtraction with whole numbers to decimal addition and subtraction, e.g., 7 tens + 8 tens = 15 tens = 150 is analogous to 7 tenths + 8 tenths = 15 tenths = 1.5.

Topic E bridges the gap between Grade 4 work with multiplication and the standard algorithm by focusing on an intermediate step—reasoning about multiplying a decimal by a one-digit whole number. The area model, with which students have had extensive experience since Grade 3, is used as a scaffold for this work.

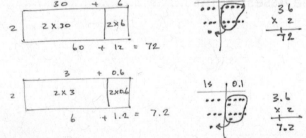

Topic F concludes Module 1 with a similar exploration of division of decimal numbers by one-digit whole number divisors. Students solidify their skills with and understanding of the algorithm before moving on to long division involving two-digit divisors in Module 2.

The mid-module assessment follows Topic C. The end-of-module assessment follows Topic F.

 | Module 1: | Place Value and Decimal Fractions
| Date: | 6/28/13

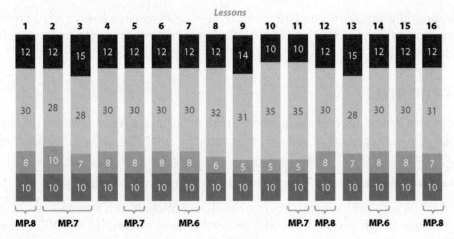

Focus Grade Level Standards

Understand the place value system.

5.NBT.1 Recognize that in a multi-digit number, a digit in one place represents 10 times as much as it represents in the place to its right and 1/10 of what it represents in the place to its left.

5.NBT.2 Explain patterns in the number of zeros of the product when multiplying a number by powers of 10, and explain patterns in the placement of the decimal point when a decimal is multiplied or divided by a power of 10. Use whole-number exponents to denote powers of 10.

5.NBT.3 Read, write, and compare decimals to thousandths.

   a. Read and write decimals to thousandths using base-ten numerals, number names, and expanded form, e.g., 347.392 = 3 × 100 + 4 × 10 + 7 × 1 + 3 × (1/10) + 9 × (1/100) + 2 × (1/1000).

   b. Compare two decimals to thousandths based on meanings of the digits in each place, using >, =, and < symbols to record the results of comparisons.

5.NBT.4 Use place value understanding to round decimals to any place.

Perform operations with multi-digit whole numbers and with decimals to hundredths.[1]

5.NBT.7 Add, subtract, multiply, and divide decimals to hundredths, using concrete models or drawings and strategies based on place value, properties of operations, and/or the relationship between addition and subtraction; relate the strategy to a written method and explain the reasoning used.

Convert like measurement units within a given measurement system.

5.MD.1 Convert among different-sized standard measurement units within a given measurement system (e.g., convert 5 cm to 0.05 m), and use these conversions in solving multi-step, real world problems.[2]

Foundational Standards

4.NBT.1 Recognize that in a multi-digit whole number, a digit in one place represents ten times what it represents in the place to its right. *For example, recognize that 700 ÷ 70 = 10 by applying concepts of place value and division.*

4.NBT.3 Use place value understanding to round multi-digit whole numbers to any place.

4.NF.5 Express a fraction with denominator 10 as an equivalent fraction with denominator 100, and use this technique to add two fractions with respective denominators 10 and 100. (Students who can generate equivalent fractions can develop strategies for adding fractions with unlike denominators in general. But addition and subtraction with unlike denominators in general is not a requirement at ths grade.) *For example, express 3/10 as 30/100, and add 3/10 + 4/100 = 34/100.*

4.NF.6 Use decimal notation for fractions with denominators 10 or 100. *For example, rewrite 0.62 as 62/100; describe a length as 0.62 meters; locate 0.62 on a number line diagram.*

4.NF.7 Compare two decimals to hundredths by reasoning about their size. Recognize that comparisons are valid only when the two decimals refer to the same whole. Record the results of comparisons with the symbols >, =, or <, and justify the conclusions, e.g., by using a visual model.

4.MD.1 Know relative sizes of measurement units within one system of units including km, m, cm; kg, g; lb, oz.; l, ml; hr, min, sec. Within a single system of measurement, express measurements in a larger unit in terms of a smaller unit. Record measurement equivalents in a two-column table. *For example, know that 1 ft is 12 times as long as 1 in. Express the length of a 4 ft snake as 48 in. Generate a conversion table for feet and inches listing the number pairs (1, 12), (2, 24), (3, 36), ...*

4.MD.2 Use the four operations to solve word problems involving distances, intervals of time, liquid volumes, masses of objects, and money, including problems involving simple fractions or

---

[1] The balance of this cluster is addressed in Module 2.
[2] The focus in this module is on the metric system to reinforce place value and writing measurements using mixed units. This standard is addressed again in later modules.

Module 1:	Place Value and Decimal Fractions
Date:	6/28/13

decimals, and problems that require expressing measurements given in a larger unit in terms of a smaller unit. Represent measurement quantities using diagrams such as number line diagrams that feature a measurement scale.

Focus Standards for Mathematical Practice

MP.6 **Attend to precision.** Students express the units of the base ten system as they work with decimal operations, expressing decompositions and compositions with understanding, e.g., "9 hundredths + 4 hundredths = 13 hundredths. I can change 10 hundredths to make 1 tenth."

MP.7 **Look for and make use of structure.** Students explore the multiplicative patterns of the base ten system when they use place value charts and disks to highlight the relationships between adjacent places. Students also use patterns to name decimal fraction numbers in expanded, unit, and word forms.

MP.8 **Look for and express regularity in repeated reasoning.** Students express regularity in repeated reasoning when they look for and use whole number general methods to add and subtract decimals and when they multiply and divide decimals by whole numbers. Students also use powers of ten to explain patterns in the placement of the decimal point and generalize their knowledge of rounding whole numbers to round decimal numbers.

Overview of Module Topics and Lesson Objectives

Standards		Topics and Objectives		Days
5.NBT.1 5.NBT.2 5.MD.1	A	**Multiplicative Patterns on the Place Value Chart**		4
		Lesson 1:	Reason concretely and pictorially using place value understanding to relate adjacent base ten units from millions to thousandths.	
		Lesson 2:	Reason abstractly using place value understanding to relate adjacent base ten units from millions to thousandths.	
		Lesson 3:	Use exponents to name place value units and explain patterns in the placement of the decimal point.	
		Lesson 4:	Use exponents to denote powers of 10 with application to metric conversions.	
5.NBT.3	B	**Decimal Fractions and Place Value Patterns**		2
		Lesson 5:	Name decimal fractions in expanded, unit, and word forms by applying place value reasoning.	
		Lesson 6:	Compare decimal fractions to the thousandths using like units and express comparisons with >, <, = .	
5.NBT.4	C	**Place Value and Rounding Decimal Fractions**		2
		Lesson 7–8:	Round a given decimal to any place using place value understanding and the vertical number line.	

Module 1: Place Value and Decimal Fractions
Date: 6/28/13

Standards		Topics and Objectives	Days
		Mid-Module Assessment: Topics A–C (assessment ½ day, return ½ day, remediation or further applications 1 day)	2
5.NBT.2 5.NBT.3 5.NBT.7	D	**Adding and Subtracting Decimals** Lesson 9: Add decimals using place value strategies and relate those strategies to a written method. Lesson 10: Subtract decimals using place value strategies and relate those strategies to a written method.	2
5.NBT.2 5.NBT.3 5.NBT.7	E	**Multiplying Decimals** Lesson 11: Multiply a decimal fraction by single-digit whole numbers, relate to a written method through application of the area model and place value understanding, and explain the reasoning used. Lesson 12: Multiply a decimal fraction by single-digit whole numbers, including using estimation to confirm the placement of the decimal point.	2
5.NBT.3 5.NBT.7	F	**Dividing Decimals** Lesson 13: Divide decimals by single-digit whole numbers involving easily identifiable multiples using place value understanding and relate to a written method. Lesson 14: Divide decimals with a remainder using place value understanding and relate to a written method. Lesson 15: Divide decimals using place value understanding including remainders in the smallest unit. Lesson 16: Solve word problems using decimal operations.	4
		End-of-Module Assessment: Topics A–F (assessment ½ day, return ½ day, remediation or further applications 1 day)	2
Total Number of Instructional Days			**20**

Module 1: Place Value and Decimal Fractions
Date: 6/28/13

Terminology

New or Recently Introduced Terms

- Thousandths (related to place value)
- Exponents (how many times a number is to be used in a multiplication sentence)
- Millimeter (a metric unit of length equal to one thousandth of a meter)
- Equation (statement that two mathematical expressions have the same value, indicated by use of the symbol =; e.g., 12 = 4 x 2 + 4)

Familiar Terms and Symbols[3]

- Centimeter (cm, a unit of measure equal to one hundredth of a meter)
- Tenths (as related to place value)
- Hundredths (as related to place value)
- Place value (the numerical value that a digit has by virtue of its position in a number)
- Base ten units (place value units)
- Digit (a numeral between 0 and 9)
- Standard form (a number written in the format: 135)
- Expanded form (e.g., 100 + 30 + 5 = 135)
- Unit form (e.g., 3.21 = 3 ones 2 tenths 1 hundredth)
- Word form (e.g., one hundred thirty-five)
- Number line (a line marked with numbers at evenly spaced intervals)
- Bundling, making, renaming, changing, regrouping, trading
- Unbundling, breaking, renaming, changing, regrouping, trading
- >, <, = (greater than, less than, equal to)
- Number sentence (e.g., 4 + 3 = 7)

Suggested Tools and Representations

- Place value charts (at least one per student for an insert in their personal board)
- Place value disks
- Number lines (a variety of templates) and a large one for the back wall of the classroom

---

[3] These are terms and symbols students have used or seen previously.

Suggested Methods of Instructional Delivery

Directions for Administration of Sprints

Sprints are designed to develop fluency. They should be fun, adrenaline-rich activities that intentionally build energy and excitement. A fast pace is essential. During Sprint administration, teachers assume the role of athletic coaches. A rousing routine fuels students' motivation to do their personal best. Student recognition of increasing success is critical, and so every improvement is celebrated.

One Sprint has two parts with closely related problems on each. Students complete the two parts of the Sprint in quick succession with the goal of improving on the second part, even if only by one more.

With practice the following routine takes about 8 minutes.

Sprint A

Pass *Sprint A* out quickly, face down on student desks with instructions to not look at the problems until the signal is given. (Some Sprints include words. If necessary, prior to starting the Sprint quickly review the words so that reading difficulty does not slow students down.)

- T: You will have 60 seconds to do as many problems as you can.
- T: I do not expect you to finish all of them. Just do as many as you can, your personal best. (If some students are likely to finish before time is up, assign a number to *count by* on the back.)
- T: Take your mark! Get set! THINK! (When you say THINK, students turn their papers over and work furiously to finish as many problems as they can in 60 seconds. Time precisely.)

After 60 seconds:

- T: Stop! Circle the last problem you did. I will read just the answers. If you got it right, call out "Yes!" and give a fist pump. If you made a mistake, circle it. Ready?
- T: (Energetically, rapid-fire call the first answer.)
- S: Yes!
- T: (Energetically, rapid-fire call the second answer.)
- S: Yes!

Repeat to the end of *Sprint A*, or until no one has any more correct. If need be, read the *count by* answers in the same way you read Sprint answers. Each number *counted by* on the back is considered a correct answer.

- T: Fantastic! Now write the number you got correct at the top of your page. This is your personal goal for Sprint B.
- T: How many of you got 1 right? (All hands should go up.)
- T: Keep your hand up until I say the number that is 1 more than the number you got right. So, if you got 14 correct, when I say 15 your hand goes down. Ready?
- T: (Quickly.) How many got 2 correct? 3? 4? 5? (Continue until all hands are down.)

Optional routine, depending on whether or not your class needs more practice with *Sprint A*:

- T: I'll give you one minute to do more problems on this half of the Sprint. If you finish, stand behind your chair. (As students work you might have the person who scored highest on *Sprint A* pass out

Module 1: Place Value and Decimal Fractions
Date: 6/28/13

Sprint B.)

T: Stop! I will read just the answers. If you got it right, call out "Yes!" and give a fist pump. If you made a mistake, circle it. Ready? (Read the answers to the first half again as students stand.)

Movement

To keep the energy and fun going, always do a stretch or a movement game in between Sprint A and B. For example, the class might do jumping jacks while skip counting by 5 for about 1 minute. Feeling invigorated, students take their seats for *Sprint B*, ready to make every effort to complete more problems this time.

Sprint B

Pass *Sprint B* out quickly, face down on student desks with instructions to not look at the problems until the signal is given. (Repeat the procedure for *Sprint A* up through the show of hands for how many right.)

T: Stand up if you got more correct on the second Sprint than on the first.

S: (Students stand.)

T: Keep standing until I say the number that tells how many more you got right on Sprint B. So if you got 3 more right on Sprint B than you did on Sprint A, when I say 3 you sit down. Ready? (Call out numbers starting with 1. Students sit as the number by which they improved is called. Celebrate the students who improved most with a cheer.)

T: Well done! Now take a moment to go back and correct your mistakes. Think about what patterns you noticed in today's Sprint.

T: How did the patterns help you get better at solving the problems?

T: Rally Robin your thinking with your partner for 1 minute. Go!

Rally Robin is a style of sharing in which partners trade information back and forth, one statement at a time per person, for about 1 minute. This is an especially valuable part of the routine for students who benefit from their friends' support to identify patterns and try new strategies.

Students may take Sprints home.

RDW or Read, Draw, Write (a Number Sentence and a Statement)

Mathematicians and teachers suggest a simple process applicable to all grades:

1) Read.
2) Draw and Label.
3) Write a number sentence (equation).
4) Write a word sentence (statement).

The more students participate in reasoning through problems with a systematic approach, the more they internalize those behaviors and thought processes.

- What do I see?
- Can I draw something?
- What conclusions can I make from my drawing?

Module 1: Place Value and Decimal Fractions
Date: 6/28/13

Modeling with Interactive Questioning	Guided Practice	Independent Practice
The teacher models the whole process with interactive questioning, some choral response, and talk moves such as "What did Monique say, everyone?" After completing the problem, students might reflect with a partner on the steps they used to solve the problem. "Students, think back on what we did to solve this problem. What did we do first?" Students might then be given the same or similar problem to solve for homework.	Each student has a copy of the question. Though guided by the teacher, they work independently at times and then come together again. Timing is important. Students might hear, "You have 2 minutes to do your drawing." Or, "Put your pencils down. Time to work together again." The Debrief might include selecting different student work to share.	The students are given a problem to solve and possibly a designated amount of time to solve it. The teacher circulates, supports, and is thinking about which student work to show to support the mathematical objectives of the lesson. When sharing student work, students are encouraged to think about the work with questions such as, "What do you see Jeremy did?" "What is the same about Jeremy's work and Sara's work?" "How did Jeremy show the 3/7 of the students?" "How did Sara show the 3/7 of the students?"

Personal Boards

Materials Needed for Personal Boards

1 High Quality Clear Sheet Protector
1 piece of stiff red tag board 11" x 8 ¼"
1 piece of stiff white tag board 11" x 8 ¼"
1 3"x 3" piece of dark synthetic cloth for an eraser
1 Low Odor Blue Dry Erase Marker: Fine Point

Directions for Creating Personal Boards

Cut your white and red tag to specifications. Slide into the sheet protector. Store your eraser on the red side. Store markers in a separate container to avoid stretching the sheet protector.

Frequently Asked Questions About Personal Boards

Why is one side red and one white?

> The white side of the board is the "paper." Students generally write on it and if working individually then turn the board over to signal to the teacher they have completed their work. The teacher then says, "Show me your boards," when most of the class is ready.

What are some of the benefits of a personal board?

- The teacher can respond quickly to a hole in student understandings and skills. "Let's do some of

Module 1: Place Value and Decimal Fractions
Date: 6/28/13

these on our personal boards until we have more mastery."
- Student can erase quickly so that they do not have to suffer the evidence of their mistake.
- They are motivating. Students love both the drill and thrill capability and the chance to do story problems with an engaging medium.
- Checking work gives the teacher instant feedback about student understanding.

What is the benefit of this personal board over a commercially purchased dry erase board?
- It is much less expensive.
- Templates such as place value charts, number bond mats, hundreds boards, and number lines can be stored between the two pieces of tag for easy access and reuse.
- Worksheets, story problems, and other problem sets can be done without marking the paper so that students can work on the problems independently at another time.
- Strips with story problems, number lines, and arrays can be inserted and still have a full piece of paper to write on.
- The red versus white side distinction clarifies your expectations. When working collaboratively, there is no need to use the red. When working independently, the students know how to keep their work private.
- The sheet protector can be removed so that student work can be projected on an overhead.

Scaffolds[4]

The scaffolds integrated into *A Story of Units* give alternatives for how students access information as well as express and demonstrate their learning. Strategically placed margin notes are provided within each lesson elaborating on the use of specific scaffolds at applicable times. They address many needs presented by English language learners, students with disabilities, students performing above grade level, and students performing below grade level. Many of the suggestions are organized by Universal Design for Learning (UDL) principles and are applicable to more than one population. To read more about the approach to differentiated instruction in *A Story of Units*, please refer to "How to Implement *A Story of Units*."

---

[4] Students with disabilities may require Braille, large print, audio, or special digital files. Please visit the website, www.p12.nysed.gov/specialed/aim, for specific information on how to obtain student materials that satisfy the National Instructional Materials Accessibility Standard (NIMAS) format.

Assessment Summary

Type	Administered	Format	Standards Addressed
Mid-Module Assessment Task	After Topic C	Constructed response with rubric	5.NBT.1 5.NBT.2 5.NBT.3 5.NBT.4 5.MD.1
End-of-Module Assessment Task	After Topic F	Constructed response with rubric	5.NBT.1 5.NBT.2 5.NBT.3 5.NBT.4 5.NBT.7 5.MD.1

COMMON CORE MATHEMATICS CURRICULUM • NY   Topic A 5•1

GRADE 5 • MODULE 1

Topic A
Multiplicative Patterns on the Place Value Chart

5.NBT.1, 5.NBT.2, 5.MD.1

Focus Standard:	5.NBT.1	Recognize that in a multi-digit number, a digit in one place represents 10 times as much as it represents in the place to its right and 1/10 of what it represents in the place to its left.
	5.NBT.2	Explain patterns in the number of zeros of the product when multiplying a number by powers of 10, and explain patterns in the placement of the decimal point when a decimal is multiplied or divided by a power of 10. Use whole-number exponents to denote powers of 10.
	5.MD.1	Convert among different-sized standard measurement units within a given measurement system (e.g., convert 5 cm to 0.05 m), and use these conversions in solving multi-step, real world problems.
Instructional Days:	4	
Coherence -Links from:	G4–M1	Place Value, Rounding, and Algorithms for Addition and Subtraction
-Links to:	G6–M2	Arithmetic Operations Including Dividing by a Fraction

Topic A begins with a conceptual exploration of the multiplicative patterns of the base ten system. This exploration extends the place value work done with multi-digit whole numbers in Grade 4 to larger multi-digit whole numbers and decimals. Students use place value disks and a place value chart to build the place value chart from millions to thousandths. Students compose and decompose units crossing the decimal with a view toward extending students' knowledge of the *10 times as large* and *1/10 as large* relationships among whole number places to that of adjacent decimal places. This concrete experience is linked to the effects on the product when multiplying any number by a power of ten. For example, students notice that multiplying 0.4 by 1000 shifts the position of the digits to the left 3 places, changing the digits' relationships to the decimal point and producing a product with a value that is 10 x 10 x 10 as large (400.0) (**5.NBT.2**). Students explain these changes in value and shifts in position in terms of place value. Additionally, students learn a new and more efficient way to represent place value units using exponents, e. g., 1 thousand = 1000 = 10^3 (**5.NBT.2**). Conversions among metric units such as kilometers, meters, and centimeters give an opportunity to apply these extended place value relationships and exponents in a meaningful context by exploring word problems in the last lesson of Topic A (**5.MD.1**).

Topic A:	Multiplicative Patterns on the Place Value Chart
Date:	6/28/13

COMMON CORE MATHEMATICS CURRICULUM • NY   Topic A 5•1

A Teaching Sequence Towards Mastery of Multiplicative Patterns on the Place Value Chart

Objective 1: Reason concretely and pictorially using place value understanding to relate adjacent base ten units from millions to thousandths.
(Lesson 1)

Objective 2: Reason abstractly using place value understanding to relate adjacent base ten units from millions to thousandths.
(Lesson 2)

Objective 3: Use exponents to name place value units and explain patterns in the placement of the decimal point.
(Lesson 3)

Objective 4: Use exponents to denote powers of 10 with application to metric conversions.
(Lesson 4)

COMMON CORE MATHEMATICS CURRICULUM • NY   Lesson 1 5•1

Lesson 1

Objective: Reason concretely and pictorially using place value understanding to relate adjacent base ten units from millions to thousandths.

Suggested Lesson Structure

- **■ Fluency Practice**   (12 minutes)
- **■ Application Problems**   (8 minutes)
- **■ Concept Development**   (30 minutes)
- **■ Student Debrief**   (10 minutes)
- **Total Time**   **(60 minutes)**

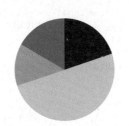

NOTES ON MULTIPLE MEANS OF ACTION AND EXPRESSION:

Throughout *A Story of Units*, place value language is key. In earlier grades, teachers use units to refer to numbers such as 245 as two *hundred* forty five. Likewise, in Grades 4 and 5, decimals should be read emphasizing their unit form. For example, 0.2 would be read 2 *tenths* rather than *zero point two*. This emphasis on unit language not only strengthens student place value understanding, but also builds important parallel between whole number and decimal fraction understanding.

Fluency Practice (12 minutes)

- Multiply by 10 **4.NBT.1**   (8 minutes)
- Rename the Units **2.NBT.1**   (2 minutes)
- Decimal Place Value **4.NF.5–6** (2 minutes)

Sprint: Multiply by 10 (8 minutes)

Materials: (S) Multiply by 10 Sprint

Note: Reviewing this fluency will acclimate students to the Sprint routine, a vital component of the fluency program.

Please see *Sprints* in the Appendix for directions on administering.

Rename the Units—Choral Response (2 minutes)

Notes: This fluency will review foundations that will lead into today's lesson.

T: (Write 10 ones = _____ ten.) Say the number sentence.
S: 10 ones = 1 ten.
T: (Write 20 ones = _____ tens.) Say the number sentence.
S: 20 ones = 2 tens.
T: 30 ones.
S: 3 tens.

Repeat the process for 80 ones, 90 ones, 100 ones, 110 ones, 120 ones, 170, 270, 670, 640, and 830.

| Lesson 1: | Reason concretely and pictorially using place value understanding to relate adjacent base ten units from millions to thousandths. |
| Date: | 6/28/13 |

1.A.3

© 2013 Common Core, Inc. All rights reserved. commoncore.org

COMMON CORE MATHEMATICS CURRICULUM • NY   Lesson 1 5•1

Decimal Place Value (2 minutes)

Materials: (S) Personal white boards

Note: Reviewing this Grade 4 topic will help lay a foundation for students to better understand place value to bigger and smaller units.

- T: (Project place value chart from millions to hundredths. Write 3 ten disks in the tens column.) How many tens do you see?
- S: 3 tens.
- T: (Write 3 underneath the disks.) There are 3 tens and how many ones?
- S: Zero ones.
- T: (Write 0 in the ones column. Below it, write 3 tens = ___.) Fill in the blank.
- S: 3 tens = 30.

Repeat the process for 3 tenths = 0.3.

- T: (Write 4 tenths = ___.) Show the answer in your place value chart.
- S: (Students write four 1 tenth disks. Below it, they write 0.4.)

Repeat the process for 3 hundredths, 43 hundredths, 5 hundredths, 35 hundredths, 7 ones 35 hundredths, 9 ones 24 hundredths, and 6 tens 2 ones 4 hundredths.

Application Problem (8 minutes)

Farmer Jim keeps 12 hens in every coop. If Farmer Jim has 20 coops, how many hens does he have in all? If every hen lays 9 eggs on Monday, how many eggs will Farmer Jim collect on Monday? Explain your reasoning using words, numbers, or pictures.

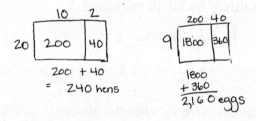

Note: This problem is intended to activate prior knowledge from Grade 4 and offer a successful start to Grade 5. Some students may use area models to solve while others may choose to use the standard algorithm. Still others may draw tape diagrams to show their thinking. Allow students to share work and compare approaches.

Concept Development (30 minutes)

Materials: (S) Personal place value mats, disks, and markers

The place value chart and its *x 10* and relationships are familiar territory for students. New learning in Grade 5 focuses on understanding a new fractional unit of *thousandths* as well as the decomposition of larger units to those that are 1/10 as large. Building the place value chart from right (tenths) to left (millions) before

Lesson 1: Reason concretely and pictorially using place value understanding to relate adjacent base ten units from millions to thousandths.
Date: 6/28/13

1.A.4

© 2013 Common Core, Inc. All rights reserved. commoncore.org

COMMON CORE MATHEMATICS CURRICULUM • NY   Lesson 1   5•1

beginning the following problem sequence may be advisable. Encourage students to multiply then bundle to form next largest place (e.g., 10 x 1 hundred = 10 hundreds, which can be bundled to form 1 thousand).

Problem 1

Divide single units by 10 to build the place value chart to introduce **thousandths**.

T: Show 1 million using disks on your chart. How can we show 1 million using hundred thousands? Work with your partner to show this on your mat.

S: 1 million is the same as 10 hundred thousands.

T: What is the result if I divide 10 hundred thousands by 10? Talk with your partner and use your mat to find the quotient.

T: (Circulate.) I saw that David put 10 disks in the hundred thousands place, then put them in 10 equal groups. How many are in each group?

S: When I divide 10 hundred thousands by 10, I get 1 hundred thousand in each group.

T: Let me record what I hear you saying. (Record on class board.)

 10 hundred thousands ÷ 10 = 1 hundred thousand
 1 million ÷ 10 = 1 hundred thousand
 1 hundred thousand is 1/10 as large as 1 million

Millions	Hundred Thousands	Ten Thousands	Thousands	Hundreds	Tens	Ones	Tenths	Hundredths	Thousandths
1	÷ 10						•		
	1						•		

T: Put 1 hundred thousand disk on your chart. What is the result if we divide 1 hundred thousand by 10? Show this on your mat and write a division sentence.

 Continue this sequence until the hundredths place is reached emphasizing the unbundling for 10 of the smaller unit and then the division. Record the place values and **equations** (using unit form) on the board being careful to point out the *1/10 as large* relationship:

MP.7

 1 million ÷ 10 = 1 hundred thousand
 1 hundred thousand ÷ 10 = 1 ten thousand
 1 ten thousand ÷ 10 = 1 thousand
 1 thousand ÷ 10 = 1 hundred
 (and so on, through 1 tenth ÷ 10 = 1 hundredth)

 NOTES ON MULTIPLE MEANS OF ENGAGEMENT:

Students who have limited experience with decimal fractions may be supported by a return to Grade 4's Module 6 to review decimal place value and symmetry with respect to the ones place.

Conversely, student understanding of decimal fraction place value units may be extended by asking for predictions of units 1/10 as large as the thousandths place, and those beyond.

Lesson 1: Reason concretely and pictorially using place value understanding to relate adjacent base ten units from millions to thousandths.
Date: 6/28/13

1.A.5

© 2013 Common Core, Inc. All rights reserved. commoncore.org

T: What patterns do you notice in the way the units are named in our place value system?

S: The ones place is the middle. There are tens on the left and tenths on the right, hundreds on the left and hundredths on the right.

T: (Point to the chart.) Using this pattern, can you predict what the name of the unit that is to the right of the hundredths place (1/10 as large as hundredths) might be? (Students share. Label the thousandths place.)

T: Thinking about the pattern that we've seen with other adjacent places, talk with your partner and predict how we might show 1 hundredth using thousandths disks and show this on your chart.

S: Just like all the other places, it takes 10 of the smaller unit to make 1 of the larger so it will take 10 thousandths to make 1 hundredth.

T: Use your chart to show the result if we divide 1 hundredth by 10 and write the division sentence. (Students share. Add this equation to the others on the board.)

NOTES ON MULTIPLE MEANS OF ENGAGEMENT:

Proportional materials such as base ten blocks can help children with language differences distinguish between place value labels like *hundredth* and *thousandth* more easily by offering clues to their relative sizes.

These students can be encouraged to name these units in their native language and then compare them to their English counterparts. Often the roots of these number words are very similar. These parallels enrich the experience and understanding of all students.

Problem 2

Multiply copies of one unit by 10, 100, and 1000.

0.4 × 10

0.04 × 10

0.004 × 10

T: Draw number disks to represent 4 tenths at the top on your place value chart.

S: (Students write.)

T: Work with your partner to find the value of 10 times 0.4. Show your result at the bottom of your place value chart.

S: 4 tenths x 10 = 40 tenths, which is the same as 4 wholes. → 4 ones is 10 times as large as 4 tenths.

T: On your place value chart, use arrows to show how the value of the digits has changed. (On place value chart, draw an arrow to indicate the shift of the digit to the left, write *x 10* above arrow.)

T: Why does the digit move one place to the left?

S: Because it is 10 times as large, it has to be bundled for the next larger unit.

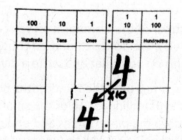

Lesson 1: Reason concretely and pictorially using place value understanding to relate adjacent base ten units from millions to thousandths.
Date: 6/28/13

Repeat with 0.03 x 10 and 0.003 x 1000. Use unit form to state each problem and encourage students to articulate how the value of the digit changes and why it changes position in the chart.

Problem 3

Divide copies of one unit by 10, 100, and 1000.

6 ÷ 10

6 ÷ 100

6 ÷ 1000

Follow similar sequence to guide students in articulating changes in value and shifts in position while showing on the place value chart.

Repeat with 0.7 ÷ 10; 0.7 ÷ 10; 0.05 ÷ 10; and 0.05 ÷ 100.

Problem 4

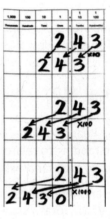

Multiply mixed units by 10, 100, and 1000.

2.43 × 10

2.43 × 100

2.43 × 1000

T: Write the digits two and forty-three hundredths on your place value chart and multiply by 10, then 100, and then 1000. Compare these products with your partner.

Lead students to discuss how the digits shift as a result in their change in value by isolating one digit, such as the 3, and comparing its value in each product.

Problem 5

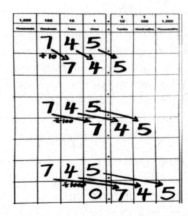

745 ÷ 10

745 ÷ 100

745 ÷ 1000

Engage in a similar discussion regarding the shift and change in value for a digit in these division problems. See discussion above.

COMMON CORE MATHEMATICS CURRICULUM • NY   Lesson 1 5•1

Problem Set (10 minutes)

Students should do their personal best to complete the Problem Set within the allotted 10 minutes. Some problems do not specify a method for solving. This is an intentional reduction of scaffolding that invokes MP.5, Use Appropriate Tools Strategically. Students should solve these problems using the RDW approach used for Application Problems.

For some classes, it may be appropriate to modify the assignment by specifying which problems students should work on first. With this option, let the careful sequencing of the problem set guide your selections so that problems continue to be scaffolded. Balance word problems with other problem types to ensure a range of practice. Assign incomplete problems for homework or at another time during the day.

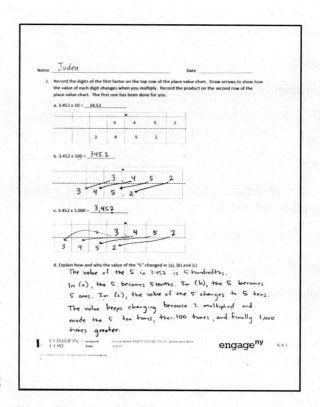

Student Debrief (10 minutes)

Lesson Objective: Reason concretely and pictorially using place value understanding to relate adjacent base ten units from millions to thousandths.

The Student Debrief is intended to invite reflection and active processing of the total lesson experience.

Invite students to review their solutions for the Problem Set. They should check work by comparing answers with a partner before going over answers as a class. Look for misconceptions or misunderstandings that can be addressed in the Debrief. Guide students in a conversation to debrief the Problem Set and process the lesson. You may choose to use any combination of the questions below to lead the discussion.

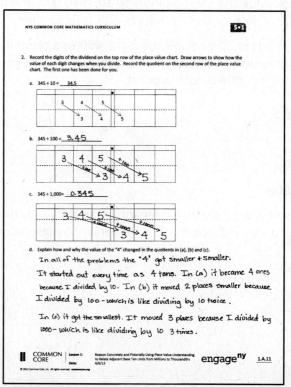

- Compare the solutions you found when multiplying by 10 and dividing by 10 (32 × 10 and 32 ÷ 10). How do the solutions in these two **equations** relate to the value of the original quantity? How do they relate to each other?
- What do you notice about the number of zeros in your products when multiplying by 10, 100, and 1000 relative to the number of places the digits shift on the place value chart? What patterns do you notice?

| Lesson 1: | Reason concretely and pictorially using place value understanding to relate adjacent base ten units from millions to thousandths. |
| Date: | 7/24/13 |

1.A.8

- What is the same and what is different about the products for Problems 1(a), 1(b), and 1(c)? (Encourage students to notice that digits are exactly the same, only the values have changed.)
- When solving Problem 2(c), many of you noticed the use of our new place value. (Lead brief class discussion to reinforce what value this place represents. Reiterate the symmetry of the places on either side of the ones place and the size of **thousandths** relative to other place values like tenths and ones.)

Exit Ticket (3 minutes)

After the Student Debrief, instruct students to complete the Exit Ticket. A review of their work will help you assess the students' understanding of the concepts that were presented in the lesson today and plan more effectively for future lessons. You may read the questions aloud to the students.

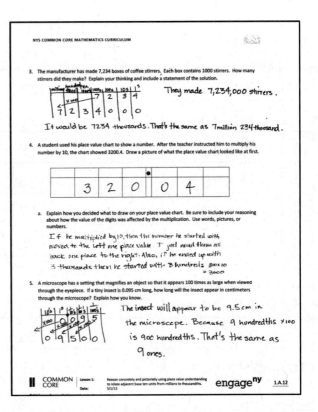

Lesson 1: Reason concretely and pictorially using place value understanding to relate adjacent base ten units from millions to thousandths.

Date: 6/28/13

A

Multiply.

Correct _____

#	Problem		#	Problem	
1	12 x 10 =		23	34 x 10 =	
2	14 x 10 =		24	134 x 10 =	
3	15 x 10 =		25	234 x 10 =	
4	17 x 10 =		26	334 x 10 =	
5	81 x 10 =		27	834 x 10 =	
6	10 x 81 =		28	10 x 834 =	
7	21 x 10 =		29	45 x 10 =	
8	22 x 10 =		30	145 x 10 =	
9	23 x 10 =		31	245 x 10 =	
10	29 x 10 =		32	345 x 10 =	
11	92 x 10 =		33	945 x 10 =	
12	10 x 92 =		34	56 x 10 =	
13	18 x 10 =		35	456 x 10 =	
14	19 x 10 =		36	556 x 10 =	
15	20 x 10 =		37	950 x 10 =	
16	30 x 10 =		38	10 x 950 =	
17	40 x 10 =		39	16 x 10 =	
18	80 x 10 =		40	10 x 60 =	
19	10 x 80 =		41	493 x 10 =	
20	10 x 50 =		42	10 x 84 =	
21	10 x 90 =		43	96 x 10 =	
22	10 x 70 =		44	10 x 580 =	

© Bill Davidson

B

Improvement _____   # Correct _____

Multiply.

#	Problem		#	Problem	
1	13 x 10 =		23	43 x 10 =	
2	14 x 10 =		24	143 x 10 =	
3	15 x 10 =		25	243 x 10 =	
4	19 x 10 =		26	343 x 10 =	
5	91 x 10 =		27	743 x 10 =	
6	10 x 91 =		28	10 x 743 =	
7	31 x 10 =		29	54 x 10 =	
8	32 x 10 =		30	154 x 10 =	
9	33 x 10 =		31	254 x 10 =	
10	38 x 10 =		32	354 x 10 =	
11	83 x 10 =		33	854 x 10 =	
12	10 x 83 =		34	65 x 10 =	
13	28 x 10 =		35	465 x 10 =	
14	29 x 10 =		36	565 x 10 =	
15	30 x 10 =		37	960 x 10 =	
16	40 x 10 =		38	10 x 960 =	
17	50 x 10 =		39	17 x 10 =	
18	90 x 10 =		40	10 x 70 =	
19	10 x 90 =		41	582 x 10 =	
20	10 x 20 =		42	10 x 73 =	
21	10 x 60 =		43	98 x 10 =	
22	10 x 80 =		44	10 x 470 =	

© Bill Davidson

Lesson 1: Reason concretely and pictorially using place value understanding to relate adjacent base ten units from millions to thousandths.

Date: 6/28/13

COMMON CORE MATHEMATICS CURRICULUM • NY   Lesson 1 Problem Set 5•1

Name _____   Date _____

1. Record the digits of the first factor on the top row of the place value chart. Draw arrows to show how the value of each digit changes when you multiply. Record the product on the second row of the place value chart. The first one has been done for you.

   a. 3.452 x 10 = ___34.52___

   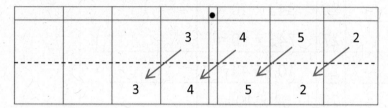

   b. 3.452 x 100 = _____

   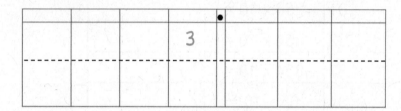

   c. 3.452 x 1000 = _____

   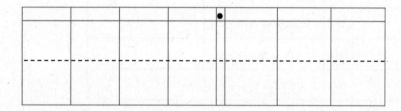

   d. Explain how and why the value of the 5 changed in (a), (b), and (c).

Lesson 1: Reason concretely and pictorially using place value understanding to relate adjacent base ten units from millions to thousandths.
Date: 6/28/13

2. Record the digits of the dividend on the top row of the place value chart. Draw arrows to show how the value of each digit changes when you divide. Record the quotient on the second row of the place value chart. The first one has been done for you.

   a. 345 ÷ 10 = ___34.5___

   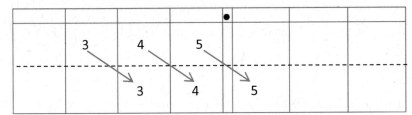

   b. 345 ÷ 100 = _____

   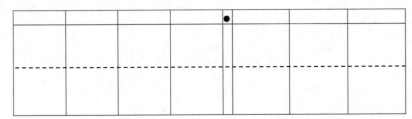

   c. 345 ÷ 1000 = _____

   d. Explain how and why the value of the 4 changed in the quotients in (a), (b), and (c).

3. A manufacturer made 7,234 boxes of coffee stirrers. Each box contains 1000 stirrers. How many stirrers did they make? Explain your thinking and include a statement of the solution.

4. A student used his place value chart to show a number. After the teacher instructed him to multiply his number by 10, the chart showed 3200.4. Draw a picture of what the place value chart looked like at first.

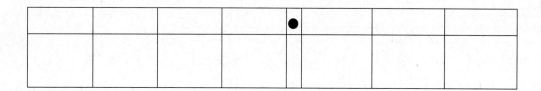

   a. Explain how you decided what to draw on your place value chart. Be sure to include your reasoning about how the value of the digits was affected by the multiplication. Use words, pictures, or numbers.

5. A microscope has a setting that magnifies an object so that it appears 100 times as large when viewed through the eyepiece. If a tiny insect is 0.095 cm long, how long will the insect appear in centimeters through the microscope? Explain how you know.

COMMON CORE MATHEMATICS CURRICULUM • NY Lesson 1 Exit Ticket 5•1

Name _____ Date _____

1. Write the first factor above the dashed line on the place value chart and the product or quotient under the dashed line, using arrows to show how the value of the digits changed. Then write your answer in the blank.

    a. 6.671 × 100 = _____

    b. 684 ÷ 1000 = _____

Lesson 1: Reason concretely and pictorially using place value understanding to relate adjacent base ten units from millions to thousandths.
Date: 6/28/13

Name _____ Date _____

1. Record the digits of the first factor on the top row of the place value chart. Draw arrows to show how the value of each digit changes when you multiply. Record the product on the second row of the place value chart. The first one has been done for you.

   a. 4.582 x 10 = ___45.82___

   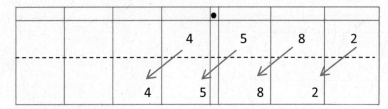

   b. 7.281 x 100 = _____

   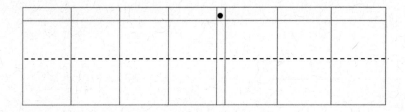

   c. 9.254 x 1000 = _____

   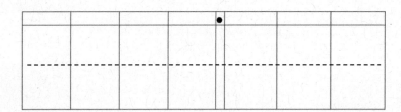

   d. Explain how and why the value of the 2 changed in (a), (b), and (c).

2. Record the digits of the dividend on the top row of the place value chart. Draw arrows to show how the value of each digit changes when you divide. Record the quotient on the second row of the place value chart. The first one has been done for you.

   a. 2.46 ÷ 10 = _____0.246_____

   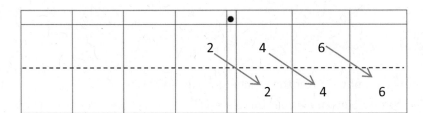

   b. 678 ÷ 100 = _____

   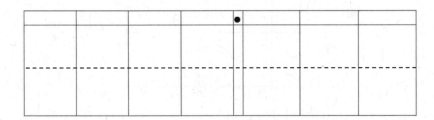

   c. 67 ÷ 1000 = _____

   d. Explain how and why the value of the 6 changed in the quotients in (a), (b), and (c).

3. Researchers counted 8,912 monarch butterflies on one branch of a tree at a site in Mexico. They estimated that the total number of butterflies at the site was 1000 times as large. About how many butterflies were at the site in all? Explain your thinking and include a statement of the solution.

4. A student used his place value chart to show a number. After the teacher instructed him to divide his number by 100, the chart showed 28.003. Draw a picture of what the place value chart looked like at first.

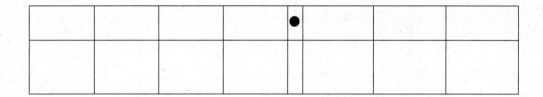

   a. Explain how you decided what to draw on your place value chart. Be sure to include your reasoning about how the value of the digits was affected by the division.

5. On a map, the perimeter of a park is 0.251 meters. The actual perimeter of the park is 1000 times as large. What is the actual perimeter of the park? Explain how you know using a place value chart.

COMMON CORE MATHEMATICS CURRICULUM • NY

Lesson 2 5•1

Lesson 2

Objective: Reason abstractly using place value understanding to relate adjacent base ten units from millions to thousandths.

Suggested Lesson Structure

- ■ Fluency Practice   (12 minutes)
- ■ Application Problems   (10 minutes)
- ■ Concept Development   (28 minutes)
- ■ Student Debrief   (10 minutes)
- **Total Time**   **(60 minutes)**

Fluency Practice (12 minutes)

- Skip-Counting **3.OA.4–6**   (3 minutes)
- Take Out the Tens **2.NBT.1**   (2 minutes)
- Bundle Ten and Change Units **4.NBT.1**   (2 minutes)
- Multiply and Divide by 10 **5.NBT.1**   (5 minutes)

NOTES ON ALIGNMENT:

Fluency tasks are included not only as warm-ups for the current lesson, but also as opportunities to retain past number understandings and to sharpen those understandings needed for coming work. Skip-counting in Grade 5 provides support for the common multiple work covered in Grade 5's Module 3.

Additionally, returning to a familiar and well understood fluency can provide a student with a feeling of success before tackling a new body of work.

Consider including body movements to accompany skip counting exercises (e.g., jumping jacks, toe touches, arm stretches, or dance movements like the Macarena).

Skip-Counting (3 minutes)

Notes: Practicing skip-counting on the number line builds a foundation for accessing higher order concepts throughout the year.

Direct the students to count forward and backward by threes to 36, emphasizing the transitions of crossing the ten. Direct the students to count forward and backward by fours to 48, emphasizing the transitions of crossing the ten.

Take Out the Tens (2 minutes)

Materials: (S) Personal white boards

Note: Decomposing whole numbers into different units will lay a foundation to do the same with decimal fractions.

T: (Write 83 ones = ____ tens ____ ones.) Write the number sentence.
S: (Students write 83 ones = 8 tens 3 ones.)

Repeat process for 93 ones, 103 ones, 113 ones, 163 ones, 263 ones, 463 ones, and 875 ones.

| Lesson 2: | Reason abstractly using place value understanding to relate adjacent base ten units from millions to thousandths. |
| Date: | 6/28/13 |

COMMON CORE MATHEMATICS CURRICULUM • NY

Lesson 2 5•1

Bundle Ten and Change Units (2 minutes)

Note: Reviewing this fluency will help students work towards mastery of changing place value units in the base ten system.

 T: (Write 10 hundreds = 1 ____.) Say the sentence, filling in the blank.
 S: 10 hundreds = 1 thousand.

Repeat the process for 10 tens = 1 ____, 10 ones = 1 ____, 10 tenths = 1 ____, 10 thousandths = 1 ____, and 10 hundredths = 1 ____.

Multiply and Divide by 10 (5 minutes)

Materials: (S) Personal white boards

Note: Reviewing this skill from Lesson 1 will help students work towards mastery.

 T: (Project place value chart from millions to thousandths.) Write three ones disks and the number below it.
 S: (Write 3 ones disks in the ones column. Below it, write 3.)
 T: Multiply by 10. Cross out each disk and the number 3 to show that you're changing its value.
 S: (Students cross out each ones disk and the 3. They draw arrows to the tens column and write 3 tens disks. Below it, they write 3 in the tens column and 0 in the ones column.)

Repeat the process for 2 hundredths, 3 tenths 2 hundredths, 3 tenths 2 hundredths 4 thousandths, 2 tenths 4 hundredths 5 thousandths, and 1 tenth 3 thousandths. Repeat the process for dividing by 10 for this possible sequence: 2 ones, 3 tenths, 2 ones 3 tenths, 2 ones 3 tenths 5 hundredths, 5 tenths 2 hundredths, and 1 ten 5 thousandths.

Application Problem (10 minutes)

A school district ordered 247 boxes of pencils. Each box contains 100 pencils. If the pencils are to be shared evenly amongst 10 classrooms, how many pencils will each class receive? Draw a place value chart to show your thinking.

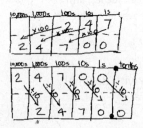

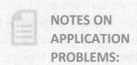

Each classroom receives 2,470 pencils.

> **NOTES ON APPLICATION PROBLEMS:**
>
> Application problems are designed to reach back to the learning in the prior day's lesson. As such, today's problem requires students to show thinking using the concrete–pictorial approach used in Lesson 1 to finding the product and quotient. This will act as an anticipatory set for today's lesson.

Lesson 2: Reason abstractly using place value understanding to relate adjacent base ten units from millions to thousandths.
Date: 6/28/13

Concept Development (28 minutes)

Materials: (S) Personal white boards

T: Turn and share with your partner. What do you remember from yesterday's lesson about how adjacent units on the place value chart are related?

S: (Students share.)

T: Moving one position to the left of the place value chart makes units 10 times larger. Conversely, moving one position to the right makes units 1 tenth the size.

As students move through the problem sets, encourage a move away from the concrete–pictorial representations of these products and quotients and a move toward reasoning about the patterns of the number of zeros in the products and quotients and the placement of the decimal.

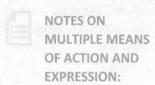

NOTES ON MULTIPLE MEANS OF ACTION AND EXPRESSION:

Although students are being encouraged toward more abstract reasoning in the lesson, it is important to keep concrete materials like place value mats and disks accessible to students while these place value relationships are being solidified. Giving students the freedom to move between levels of abstraction on a task by task basis can decrease anxiety when working with more difficult applications.

Problems 1–4

367 × 10

367 ÷ 10

4,367 × 10

4,367 ÷ 10

T: Work with your partner to solve these problems. Write two complete number sentences on your board.

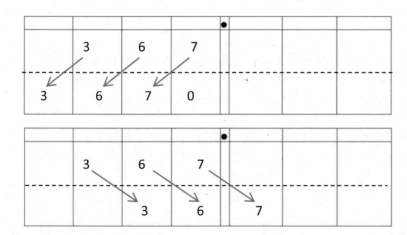

S: 367 × 10 = 3670. 367 ÷ 10 = 36.7

COMMON CORE MATHEMATICS CURRICULUM • NY   Lesson 2 5•1

MP.3

T: Explain how you got your answers. What are the similarities and differences between the two answers?

S: Digits are the same but their values have changed so their position in the number is different. → The 3 is 10 times larger in the product than in the factor. It was 3 hundreds, now it is 3 thousands. → The six started out as 6 tens, but once it was divided by 10, it is now 1 tenth as large because it is 6 ones.

MP.2

T: What patterns do you notice in the number of zeros in the product and the placement of the decimal in the quotient? What do you notice about the number of zeros in your factors and the shift in places in you product? What do you notice about the number of zeros in your divisor and shift in places in your quotient?

S: (Students share.)

Repeat this sequence with the last pair of expressions. Encourage students with this pair to visualize the place value mat and attempt to find the product and quotient without drawing the mat. Circulate watching for misconceptions and students who are not ready to work on an abstract level. As students share thinking encourage the use of the language 10 times as large and 1/10 as large.

Problems 5–8

215.6 × 100

215.6 ÷ 100

3.7 × 100

3.7 ÷ 100

T: Now solve with your partner by visualizing your place value mat and recording only your products and quotients. You may check your work using a place value mat. (Circulate, looking for students who may still need the support of the place value mat.)

S: (Students solve.)

MP.7

T: Compare your work with your partner's. Do you agree? How many times did the digit shift in each problem and why? Share your thinking with your partner.

S: The digits shifted two places to the left when we multiply and shifted two places to the right when we divide. → This time the numbers each shifted 2 places because there are 2 zeros in 100. → The values of the products are 100 times as large, so the digits had to shift to larger units.

Problems 9–10

0.482 × 1000

482 ÷ 1000

Follow a similar sequence for these equations.

Lesson 2: Reason abstractly using place value understanding to relate adjacent base ten units from millions to thousandths.
Date: 6/28/13

Lesson 2 5•1

Problem Set (10 minutes)

Students should do their personal best to complete the Problem Set within the allotted 10 minutes. For some classes, it may be appropriate to modify the assignment by specifying which problems they work on first. Some problems do not specify a method for solving. Students solve these problems using the RDW approach used for Application Problems.

Student Debrief (10 minutes)

Lesson Objective: Reason abstractly using place value understanding to relate adjacent base ten units from millions to thousandths.

The Student Debrief is intended to invite reflection and active processing of the total lesson experience.

Invite students to review their solutions for the Problem Set. They should check work by comparing answers with a partner before going over answers as a class. Look for misconceptions or misunderstandings that can be addressed in the Debrief. Guide students in a conversation to debrief the Problem Set and process the lesson. You may choose to use any combination of the questions below to lead the discussion.

- Compare and contrast answers in Problem 1(a) and (b), or (c) and (d)?
- What's similar about the process you used to solve Problem 1(a), (c), (e), and (g)?
- What's similar about the process you used to solve Problem 1(b), (d), (f), and (h)?
- When asked to find the number 1 tenth as large as another number, what operation would you use? Explain how you know.
- When solving Problem 2, how did the number of zeros in the factors help you determine the product?
- Can you think of a time when there will be a different number of zeros in the factors and the product? (If students have difficulty answering, give them the example of 4 x 5, 4 x 50, 40 x 50. Then ask if they can think of other examples.)

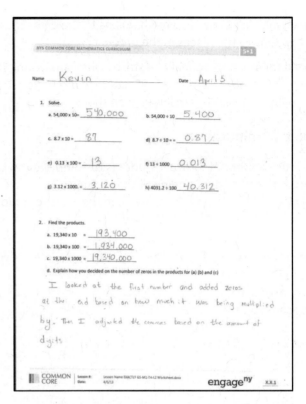

Lesson 2: Reason abstractly using place value understanding to relate adjacent base ten units from millions to thousandths.
Date: 6/28/13

- When dividing by 10, what happens to the digits in the quotient? What multiplying by 100, what happens to the places in the product?

Be prepared for students to make mistakes when answering Problem 4. (Using a place value chart to solve this problem may reduce the errors. Encourage discussion about the relative size of the units in relation to a whole and why hundredths are larger than thousandths.)

Exit Ticket (3 minutes)

After the Student Debrief, instruct students to complete the Exit Ticket. A review of their work will help you assess the students' understanding of the concepts that were presented in the lesson today and plan more effectively for future lessons. You may read the questions aloud to the students.

Lesson 2: Reason abstractly using place value understanding to relate adjacent base ten units from millions to thousandths.
Date: 6/28/13

COMMON CORE MATHEMATICS CURRICULUM • NY **Lesson 2 Problem Set 5•1**

Name _____ Date _____

1. Solve.

    a. 54,000 x 10 = _____   e. 0.13 x 100 = _____

    b. 54,000 ÷ 10 = _____   f. 13 ÷ 1000 = _____

    c. 8.7 x 10 = _____   g. 3.12 x 1000 = _____

    d. 8.7 ÷ 10 = _____   h. 4031.2 ÷ 100 = _____

2. Find the products.

    a. 19,340 x 10   = _____

    b. 19,340 x 100   = _____

    c. 19,340 x 1000 = _____

    d. Explain how you decided on the number of zeros in the products for (a), (b), and (c).

3. Find the quotients.

    a. 152 ÷ 10 = _____

    b. 152 ÷ 100 = _____

    c. 152 ÷ 1000 = _____

    d. Explain how you decided where to place the decimal in the quotients in (a), (b), and (c).

4. Janice thinks that 20 hundredths is equivalent to 2 thousandths because 20 hundreds is equal to 2 thousands. Use words and a place value chart to correct Janice's error.

5. Canada has a population that is about 1/10 as large as the United States. If Canada's population is about 32 million, about how many people live in the United States? Explain the number of zeros in your answer.

COMMON CORE MATHEMATICS CURRICULUM • NY   Lesson 2 Exit Ticket   5•1

Name _____   Date _____

1. Solve.

    a. 32.1 x 10 = _____   b. 3632.1 ÷ 10 = _____

2. Solve.

    a. 455 x 1000 = _____   b. 455 ÷ 1000 = _____

Name _____ Date _____

1. Solve.

    a. 36,000 x 10 = _____

    b. 36,000 ÷ 10 = _____

    c. 4.3 x 10 = _____

    d. 4.3 ÷ 10 = _____

    e. 0.24 x 100 = _____

    f. 24 ÷ 1000 = _____

    g. 4.54 x 1000 = _____

    h. 3045.4 ÷ 100 = _____

2. Find the products.

    a. 14,560 x 10   = _____

    b. 14,560 x 100   = _____

    c. 14,560 x 1000 = _____

    d. Explain how you decided on the number of zeros in the products for (a), (b), and (c).

3. Find the quotients.

    a. 1.65 ÷ 10 = _____

    b. 1.65 ÷ 100 = _____

    c. Explain how you decided where to place the decimal in the quotients in (a), (b), and (c).

4. Ted says that 3 tenths multiplied by 100 equal 300 thousandths. Is he correct? Use a place value chart to explain your answer.

5. Alaska has a land area of about 1,700,000 km². Florida has a land area 1/10 the size of Alaska. What is the land area of Florida? Explain how you found your answer.

Lesson 3

Objective: Use exponents to name place value units and explain patterns in the placement of the decimal point.

Suggested Lesson Structure

- **Fluency Practice**   (15 minutes)
- **Application Problems**   (7 minutes)
- **Concept Development**   (28 minutes)
- **Student Debrief**   (10 minutes)
- **Total Time**   **(60 minutes)**

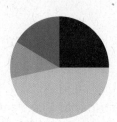

Fluency Practice (15 minutes)

- Multiply by 3 **3.OA.7**   (8 minutes)
- State the Unit as a Decimal **5.NBT.2**   (4 minutes)
- Multiply by 10, 100, and 1000 **5.NBT.2**   (3 minutes)

Sprint: Multiply by 3 (8 minutes)

Materials: (S) Multiply by 3 Sprint.

Notes: This fluency will review foundational skills learned in Grades 3 and 4.

State the Unit as a Decimal—Choral Response (4 minutes)

Notes: Reviewing these skills will help students work towards mastery of decimal place value, which will help them apply their place value skills to more difficult concepts.

T: (Write 9 tenths = _____.)
S: 0.9
T: (Write 10 tenths = _____.)
S: 1.0
T: Write 11 tenths = _____.)
S: 1.1
T: (Write 12 tenths = _____.)
S: 1.2
T: (Write 18 tenths = _____.)

S: 1.8
T: (Write 28 tenths = ____.)
S: 2.8
T: (Write 58 tenths = ____.)
S: 5.8

Repeat the process for 9 hundredths, 10 hundredths, 20 hundredths, 60 hundredths, 65 hundredths, 87 hundredths, and 118 tenths. (This last item is an extension.)

Multiply and Divide by 10, 100, and 1000 (3 minutes)

Materials: (S) Personal white boards.

Notes: This fluency drill will review concepts taught in Lesson 2.

T: (Project place value chart from millions to thousandths.) Write two 1 thousandths disks and the number below it.
S: (Students write two 1 thousandths disks in the thousandths column. Below it, they write 0.002.)
T: Multiply by 10. Cross out each disk and the number 2 to show that you're changing its value.
S: (Students cross out each 1 thousandths disk and the 2. They draw arrows to the hundredths column and write two 1 hundredth disks. Below it, they write 2 in the hundredths column and 0 in the ones and tenths column.)

Repeat the process for the possible sequence: 0.004 x 100; 0.004 x 1000; 1.004 x 1000; 1.024 x 100; 1.324 x 100; 1.324 x 10; and 1.324 x 1000.

Repeat the process for dividing by 10, 100, and 1000 for this possible sequence: 4 ÷ 10; 4.1 ÷ 10; 4.1 ÷ 100; 41 ÷ 1000; and 123 ÷ 1000.

NOTES ON MULTIPLE MEANS OF ACTION AND EXPRESSION:

Very large numbers like *one million* and beyond easily capture the imagination of students. Consider allowing students to research and present to classmates the origin of number names like *googol* and *googleplex*. Connections to literacy can also be made with books about large numbers, such as *How Much is a Million* by Steven Kellogg, *A Million Dots* by Andrew Clements, *Big Numbers and Pictures That Show Just How Big They Are* by Edward Packard and Sal Murdocca.

The following benchmarks may help students appreciate just how large a *googol* is.

- There are approximately 10^{24} stars in the observable universe.
- There are approximately 10^{80} atoms in the observable universe.
- A stack of 70 numbered cards can be ordered in approximately 1 *googol* different ways. That means that that the number of ways a stack of only 70 cards can be shuffled is more than the number of atoms in the observable universe.

Application Problem (7 minutes)

Jack and Kevin are creating a mosaic by using fragments of broken tiles for art class. They want the mosaic to have 100 sections. If each section requires 31.5 tiles, how many tiles will they need to complete the mosaic? Explain your reasoning with a place value chart.

COMMON CORE MATHEMATICS CURRICULUM • NY   Lesson 3   5•1

Concept Development (28 minutes)

Materials: (S) Personal white boards

Problem 1

T: (Draw or project chart, adding numerals as discussion unfolds.)

				100	10
				10 x 10	10 x 1

T: (Write 10 × ____ = 10 on the board.) On your personal board, fill in the missing factor to complete this number sentence.

S: (Students write.)

T: (Write 10 × ____ = 100 on the board.) Fill in the missing factor to complete this number sentence.

S: (Students write.)

T: This time, using only 10 as a factor, how could you multiply to get a product of 1000? Write the multiplication sentence on your personal board.

S: 10 x 10 x 10 = 1000.

T: Work with your partner. What would the multiplication sentence be for 10,000 using only 10 as a factor? Write on your personal board.

S: (Students write.)

MP.7

T: How many factors of 10 did we have to multiply to get to 1000?

S: 3.

T: How many factors of 10 do we have to multiply to get 10,000?

S: 4.

T: Say the number sentence.

S: 10 x 10 x 10 x 10 = 10,000.

T: How many zeros are in our product, 10,000?

S: 4 zeros.

T: What patterns do you notice? Turn and share with your partner.

S: The number of zeros is the same on both side of the equation. → The number of zeros in the product is the same as the number of zeros in the factors. → I see three zeros on the left side, and there are three zeros on the right side for 10 x 10 x 10 = 1000. → The 1 moves one place to the left every time we multiply by 10. → It's like a place value chart. Each number is 10 times as much as the last one.

Lesson 3:   Use exponents to name place value units and explain patterns in
Date: the placement of the decimal point.
             6/28/13

1.A.32

© 2013 Common Core, Inc. All rights reserved. commoncore.org

T: Using this pattern, how many factors of 10 do we have to multiply to get 1 million? Work with your partner to write the multiplication sentence.

S: (Students write.)

T: How many factors of 10 did you use?

S: 6

T: Why did we need 6 factors of 10?

S: 1 million has 6 zeros.

T: We can use an **exponent** (write term on the board) to represent how many times we use 10 as a factor. We can write 10 x 10 as 10^2. (Add to the chart.) We say, "Ten to the second power." The *2* (point to exponent) is the exponent and it tells us how many times to use 10 as a factor.

T: How do you express 1000 using exponents? Turn and share with your partner.

S: We multiply 10 × 10 × 10, that's three times, so the answer is 10^3. → There are three zeros in 1000, so it's ten to the third power.

T: Working with your partner, complete the chart using the exponents to represent the each value on the place value chart.

1,000,000	100,000	10,000	1000	100	10
(10 x 10 x 10) x (10 x 10 x 10)	10x 10 x (10 x 10 x 10)	10 x (10 x 10 x 10)	(10 x 10 x 10)	10 x 10	10 x 1
10^6	10^5	10^4	10^3	10^2	10^1

After reviewing the chart with the students, challenge them to multiply 10 one hundred times. As some start to write it out, others may write 10^{100}, a *googol*, with exponents.

T: Now look at the place value chart; let's read our powers of 10 and the equivalent values.

S: Ten to the second power equals 100; ten to the third power equals 1000. (Continue to read chorally up to 1 million.)

T: Since a *googol* has 100 zeros, write it using an exponent on your personal board.

S: (Students write 10^{100}.)

Problem 2

10^5

T: Write *ten to the fifth power* as a product of tens.

S: 10^5 = 10 x 10 x 10 x 10 x 10.

T: Find the product.

S: 10^5 = 100,000.

Repeat with more examples as needed.

NOTES ON MULTIPLE MEANS OF REPRESENTATIONS:

Providing non-examples is a powerful way to clear up mathematical misconceptions and generate conversation around the work. Highlight those examples such as 10^5 pointing out its equality to 10 x 10 x 10 x 10 x 10 but not to 10 x 5 or even 5^{10}.

Allowing students to explore a calculator and highlighting the functions used to calculate these expressions (e.g., 10^5 versus 10 x 5) can be valuable.

Lesson 3: Use exponents to name place value units and explain patterns in the placement of the decimal point.
Date: 6/28/13

Problem 3

10 x 100

T: Work with your partner to write this expression using an exponent on your personal board. Explain your reasoning.

S: I multiply 10 x 100 to get 1000, so the answer is ten to the third power. → There are 3 factors of 10. → There are three 10's. I can see one 10 in the first factor and 2 more tens in the second factor.

Repeat with 100 x 1000 and other examples as needed.

Problems 4–5

3×10^2

3.4×10^3

T: Compare this expression to the ones we've already talked about.

S: These have factors other than 10.

T: Write 3×10^2 without using an exponent. Write it on your personal board.

S: 3 x 100.

T: What's the product?

S: 300.

T: If you know that 3 × 100 is 300, then what is 3×10^2? Turn and explain to your partner.

S: The product is also 300. 10^2 and 100 are same amount so the product will be the same.

T: Use what you learned about multiplying decimals by 10, 100, and 100 and your new knowledge about exponents to solve 3.4×10^3 with your partner.

S: (Students work.)

Repeat with 4.021×10^2 and other examples as needed.

Have students share their solutions and reasoning about multiplying decimal factors by powers of ten. In particular, students should articulate the relationship between the exponent and how the values of the digits change and placement of the decimal in the product.

Problems 6–7

$700 \div 10^2$

$7.1 \div 10^2$

T: Write $700 \div 10^2$ without using an exponent and find the quotient. Write it on your personal board.

S: 700 ÷ 100 = 7

T: If you know that 700 ÷ 100 is 7, then what is $700 \div 10^2$? Turn and explain to your partner.

S: The quotient is 7 because $10^2 = 100$.

T: Use what you know about dividing decimals by multiples of 10 and your new knowledge about exponents to solve $7.1 \div 10^2$ with your partner.

S: (Students work.)

Lesson 3: Use exponents to name place value units and explain patterns in the placement of the decimal point.
Date: 6/28/13

T: Tell your partner what you notice about the relationship between the exponents and how the values of the digits change. Also discuss how you decided where to place the decimal.

Repeat with more examples as needed.

Problems 8–9

Complete this pattern: 0.043   4.3   430   _____   _____   _____

T: (Write the pattern on the board.) Turn and talk with your partner about the pattern on the board. How is the value of the 4 changing as we move to the next term in the sequence? Draw a place value chart to explain your ideas as you complete the pattern and use an exponent to express the relationships.

S: The 4 moved two places to the left. → Each number is being multiplied by 100 to get the next one. → Each number is multiplied by 10 twice. → Each number is multiplied by 10^2.

Repeat with 6,300,000; ____; 630; 6.3; _____ and other patterns as needed.

T: As you work on the Problem Set, be sure you are thinking about the patterns that we've discovered today.

Problem Set (10 minutes)

Students should do their personal best to complete the Problem Set within the allotted 10 minutes. For some classes, it may be appropriate to modify the assignment by specifying which problems they work on first. Some problems do not specify a method for solving. Students solve these problems using the RDW approach used for Application Problems.

Student Debrief (10 minutes)

Lesson Objective: Use exponents to name place value units and explain patterns in the placement of the decimal point.

The Student Debrief is intended to invite reflection and active processing of the total lesson experience.

Invite students to review their solutions for the Problem Set. They should check work by comparing answers with a partner before going over answers as a class. Look for misconceptions or misunderstandings that can be addressed in the Debrief. Guide students in a conversation to debrief the Problem Set and process the lesson. You may choose to use any combination of the questions below to lead the discussion.

- What is an **exponent** and how can exponents be useful in representing numbers? (This question could also serve as a prompt for math journals. Journaling about new vocabulary throughout the year can be a powerful way for students to solidify their understanding of new terms.)
- How would you write 1000 using exponents? How would you write it as a multiplication sentence using only 10 as a factor?
- Explain to your partner the relationship we saw between the exponents and the number of the places the digit shifted when you multiply or divide by a power of 10.
- How are the patterns you discovered in Problem 3 and 4 in the Problem Set alike?

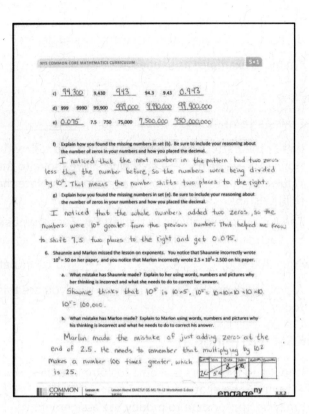

Give students plenty of opportunity to discuss the error patterns in Problem 6(a) and 6(b). These are the most common misconceptions students hold when dealing with exponents, so it is worth the time to see that they do not become firmly held.

Exit Ticket (3 minutes)

After the Student Debrief, instruct students to complete the Exit Ticket. A review of their work will help you assess the students' understanding of the concepts that were presented in the lesson today and plan more effectively for future lessons. You may read the questions aloud to the students.

Lesson 3: Use exponents to name place value units and explain patterns in the placement of the decimal point.
Date: 6/28/13

A Multiply.

Correct _____

1	1 x 3 =		23	10 x 3 =	
2	3 x 1 =		24	9 x 3 =	
3	2 x 3 =		25	4 x 3 =	
4	3 x 2 =		26	8 x 3 =	
5	3 x 3 =		27	5 x 3 =	
6	4 x 3 =		28	7 x 3 =	
7	3 x 4 =		29	6 x 3 =	
8	5 x 3 =		30	3 x 10 =	
9	3 x 5 =		31	3 x 5 =	
10	6 x 3 =		32	3 x 6 =	
11	3 x 6 =		33	3 x 1 =	
12	7 x 3 =		34	3 x 9 =	
13	3 x 7 =		35	3 x 4 =	
14	8 x 3 =		36	3 x 3 =	
15	3 x 8 =		37	3 x 2 =	
16	9 x 3 =		38	3 x 7 =	
17	3 x 9 =		39	3 x 8 =	
18	10 x 3 =		40	11 x 3 =	
19	3 x 10 =		41	3 x 11 =	
20	3 x 3 =		42	12 x 3 =	
21	1 x 3 =		43	3 x 13 =	
22	2 x 3 =		44	13 x 3 =	

© Bill Davidson

Lesson 3: Use exponents to name place value units and explain patterns in the placement of the decimal point.

Date: 6/28/13

B   Multiply.   Improvement _____   # Correct _____

#	Problem	Answer	#	Problem	Answer
1	3 x 1 =		23	9 x 3 =	
2	1 x 3 =		24	3 x 3 =	
3	3 x 2 =		25	8 x 3 =	
4	2 x 3 =		26	4 x 3 =	
5	3 x 3 =		27	7 x 3 =	
6	3 x 4 =		28	5 x 3 =	
7	4 x 3 =		29	6 x 3 =	
8	3 x 5 =		30	3 x 5 =	
9	5 x 3 =		31	3 x 10 =	
10	3 x 6 =		32	3 x 1 =	
11	6 x 3 =		33	3 x 6 =	
12	3 x 7 =		34	3 x 4 =	
13	7 x 3 =		35	3 x 9 =	
14	3 x 8 =		36	3 x 2 =	
15	8 x 3 =		37	3 x 7 =	
16	3 x 9 =		38	3 x 3 =	
17	9 x 3 =		39	3 x 8 =	
18	3 x 10 =		40	11 x 3 =	
19	10 x 3 =		41	3 x 11 =	
20	1 x 3 =		42	13 x 3 =	
21	10 x 3 =		43	3 x 13 =	
22	2 x 3 =		44	12 x 3 =	

© Bill Davidson

Lesson 3: Use exponents to name place value units and explain patterns in the placement of the decimal point.
Date: 6/28/13

Name _____   Date _____

1. Write the following in exponential form (e.g., $100 = 10^2$).

   a. 10,000 = _____

   b. 1000 = _____

   c. 10 × 10 = _____

   d. 100 × 100 = _____

   e. 1,000,000 = _____

   f. 1000 × 1000 = _____

2. Write the following in standard form (e.g., $5 \times 10^2 = 500$).

   a. 9×10^3 = _____

   b. 39×10^4 = _____

   c. $7200 \div 10^2$ = _____

   d. $7,200,000 \div 10^3$ = _____

   e. 4.025×10^3 = _____

   f. 40.25×10^4 = _____

   g. $725 \div 10^3$ = _____

   h. $7.2 \div 10^2$ = _____

3. Think about the answers to Problem 2(a–d). Explain the pattern used to find an answer when you multiply or divide a whole number by a power of 10.

4. Think about the answers to Problem 2(e–h). Explain the pattern used to place the decimal in the answer when you multiply or divide a decimal by a power of 10.

5. Complete the patterns.

   a. 0.03   0.3   _____   30   _____   _____

   b. 6,500,000   65,000   _____   6.5   _____

   c. _____   9,430   _____   94.3   9.43   _____

   d. 999   9990   99,900   _____   _____   _____

   e. _____   7.5   750   75,000   _____   _____

   f. Explain how you found the missing numbers in set (b). Be sure to include your reasoning about the number of zeros in your numbers and how you placed the decimal.

   g. Explain how you found the missing numbers in set (d). Be sure to include your reasoning about the number of zeros in your numbers and how you placed the decimal.

6. Shaunnie and Marlon missed the lesson on exponents. Shaunnie incorrectly wrote $10^5 = 50$ on her paper, and Marlon incorrectly wrote $2.5 \times 10^2 = 2.500$ on his paper.

   a. What mistake has Shaunnie made? Explain using words, numbers, and pictures why her thinking is incorrect and what she needs to do to correct her answer.

   b. What mistake has Marlon made? Explain using words, numbers, and pictures why his thinking is incorrect and what he needs to do to correct his answer.

Lesson 3: Use exponents to name place value units and explain patterns in the placement of the decimal point.
Date: 6/28/13

Name _____ Date _____

1. Write the following in exponential form and as a multiplication sentence using only 10 as a factor (e.g., $100 = 10^2 = 10 \times 10$).

    a. 1,000   = _____ = _____

    b. 100 × 100   = _____ = _____

2. Write the following in standard form (e.g., $4 \times 10^2 = 400$).

    a. 3×10^2 = _____

    b. 2.16×10^4 = _____

    c. $800 \div 10^2$ = _____

    d. $754.2 \div 10^3$ = _____

COMMON CORE MATHEMATICS CURRICULUM • NY   Lesson 3 Homework   5•1

Name _____   Date _____

1. Write the following in exponential form (e.g., $100 = 10^2$).

   a. 1000 = _____   d. 100 x 10 = _____

   b. 10 × 10 = _____   e. 1,000,000 = _____

   c. 100,000 = _____   f. 10,000 × 10 = _____

2. Write the following in standard form (e.g., $4 \times 10^2 = 400$).

   a. 4×10^3 = _____   e. 6.072×10^3 = _____

   b. 64×10^4 = _____   f. 60.72×10^4 = _____

   c. $5300 \div 10^2$ = _____   g. $948 \div 10^3$ = _____

   d. $5,300,000 \div 10^3$ = _____   h. $9.4 \div 10^2$ = _____

3. Complete the patterns.

   a. 0.02   0.2   _____   20   _____   _____

   b. 3,400,000   34,000   _____   3.4   _____

   c. _____   8,570   _____   85.7   8.57   _____

   d. 444   4440   44,400   _____   _____   _____

   e. _____   9.5   950   95,000   _____   _____

Lesson 3: Use exponents to name place value units and explain patterns in the placement of the decimal point.
Date: 6/28/13

1.A.42

4. After a lesson on exponents, Tia went home and said to her mom, "I learned that 10^4 is the same as 40,000." She has made a mistake in her thinking. Use words, numbers or a place value chart to help Tia correct her mistake.

5. Solve $247 \div 10^2$ and 247×10^2.

   a. What is different about the two answers? Use words, numbers or pictures to explain how the decimal point shifts.

   b. Based on the answers from the pair of expressions above, solve $247 \div 10^3$ and 247×10^3.

COMMON CORE MATHEMATICS CURRICULUM • NY   Lesson 4 5•1

Lesson 4

Objective: Use exponents to denote powers of 10 with application to metric conversions.

Suggested Lesson Structure

■ Fluency Practice   (12 minutes)
■ Application Problems   (8 minutes)
■ Concept Development   (30 minutes)
■ Student Debrief   (10 minutes)
  Total Time   **(60 minutes)**

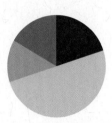

Fluency Practice (12 minutes)

- Multiply and Divide Decimals by 10, 100,
  and 1000 **5.NBT.2**   (5 minutes)
- Write the Unit as a Decimal **5.NBT.1**   (2 minutes)
- Write in Exponential Form **5.NBT.2**   (3 minutes)
- Convert Units **4.MD.1**   (2 minutes)

Multiply and Divide Decimals by 10, 100, and 1000 (5 minutes)

Materials: (S) Personal white boards

Note: This fluency drill will review concepts taught in earlier lessons and help students work towards mastery in multiplying and dividing decimals by 10, 100, and 1000.

T: (Project place value chart from millions to thousandths. Write 3 disks in the tens column, 2 disks in the ones column, and 4 disks in the tenths column.) Say the value as a decimal.
S: 32.4 (thirty-two and four tenths).
T: Write the number on your personal boards and multiply it by ten.

Students write 32.4 on their place value charts, cross out each digit, and shift the number one place value to the left to show 324.

T: Show 32.4 divided by 10.

Students write 32.4 on their place value charts, cross out each digit, and shift the number one place value to the right to show 3.24.

Repeat the process and sequence for 32.4 x 100; 32.4 ÷ 100; 837 ÷ 1000; and 0.418 x 1000.

   Lesson 4: Use exponents to denote powers of 10 with application to metric conversions.
         Date: 6/28/13

COMMON CORE MATHEMATICS CURRICULUM • NY   Lesson 4 5•1

Write the Unit as a Decimal (2 minutes)

Materials: (S) Personal white boards

Note: Reviewing these skills will help students work towards mastery of decimal place value, which will in turn help them apply their place value skills to more difficult concepts.

T: 9 tenths.
S: 0.9
T: 10 tenths.
S: 1.0

Repeat the process for 20 tenths, 30 tenths, 70 tenths, 9 hundredths, 10 hundredths, 11 hundredths, 17 hundredths, 57 hundredths, 42 hundredths, 9 thousandths, 10 thousandths, 20 thousandths, 60 thousandths, 64 thousandths, and 83 thousandths.

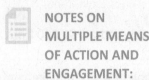

NOTES ON MULTIPLE MEANS OF ACTION AND ENGAGEMENT:

Consider posting a class-size place value chart as an aid to students in visualizing the unit work of this fluency activity.

It may also be fruitful to have students verbalize their reasoning about the equivalence of 10 tenths to 1.0, 20 tenths to 2.0, etc.

Write in Exponential Form (3 minutes)

Materials: (S) Personal white boards

Note: Reviewing this skill in isolation will lay a foundation for students to apply the skill in multiplication during the lesson.

T: (Write 100 = $10^?$.) Write 100 in exponential form.
S: (Students write 100 = 10^2.)

Repeat the process for 1000, 10,000, and 1,000,000.

Convert Units (2 minutes)

Materials: (S) Personal white boards

Note: Reviewing conversions in isolation will lay a foundation for students to apply this knowledge through multiplication and division during the lesson.

Use this quick fluency to activate prior knowledge of these familiar equivalents.

T: (Write 1 km = _____ m.) Fill in the missing number.
S: (Students write 1 km = 1000 m.)

Repeat process and procedure for 1 kg = _____ g, 1 liter = _____ ml, 1 m = _____ cm.

| Lesson 4: | Use exponents to denote powers of 10 with application to metric conversions. |
| Date: | 6/28/13 |

1.A.45

COMMON CORE MATHEMATICS CURRICULUM • NY   Lesson 4   5•1

Application Problem (8 minutes)

Mr. Brown wants to withdraw $1,000 from his bank and in ten dollar bills. How many ten dollar bills should he receive?

Note: Use this problem with a familiar context of money to help students begin to use various units to rename the same quantity—the focus of today's lesson.

Concept Development (30 minutes)

Materials: (S) Meter strip, markers

Each problem set below includes conversions both from larger to smaller units and smaller to larger units. Allow students the time to reason about how the change in the size of the unit will affect the *quantity* of units needed to express an equivalent measure rather than giving rules about whether to multiply or divide.

Problem 1

Draw a line that is 2 meters long and convert it to centimeters and **millimeters**.

> **NOTES ON MULTIPLE MEANS OF ACTION AND ENGAGEMENT:**
>
> As discussions ensue about conversions from meters to kilometers, centimeters and millimeters, take the opportunity to extend thinking by asking students to make a conversion to the unit that is 1/10 as large as a meter (decimeter) and the unit 10 times as large (decameter). Students can make predictions about the names of these units or do research about these and other metric units that are less commonly used. Students might also make connections to real world mathematics by investigating industry applications for the less familiar units.

MP.3

T: Draw a line 2 meters long.
S: (Students draw.)
T: With your partner, determine how many centimeters equal 2 meters.
S: 200 centimeters.
T: How is it that the same line can measure both 2 meters and 200 centimeters?
T: Discuss with a partner how we convert from 2 meters to 200 centimeters?
S: (After talking with a partner.) Multiply by 100.
T: Why didn't the length change? Discuss that with your partner.

Repeat the same sequence with millimeters.

T: Can we represent the conversion from meters to centimeters or meters to millimeters with exponents? Discuss this with your partner.

Let them see that to convert to centimeters from meters, we multiplied by 10^2, while to convert from meters to millimeters we multiplied by 10^3. Repeat the same sequence in reverse so that students see that to convert from centimeters to meters we divide by 10^2 and to convert from millimeters to meters we divide by 10^3. If there seems to be a large lack of clarity do another conversion with 1 meter or 3 meters.

COMMON CORE

Lesson 4: Use exponents to denote powers of 10 with application to metric conversions.
Date: 6/28/13

Problem 2

Convert 1.37 meters to centimeters and millimeters.

- T: Draw a line 1 meter 37 centimeters long.
- S: (Students draw.)
- T: What fraction of a whole meter is 37 centimeters?
- S: 37 hundredths.
- T: Write 1 and 37 hundredths as a decimal fraction.
- T: With your partner, determine how many centimeters is equal to 1.37 meters both by looking at your meter strip and line and writing an equation using an exponent.
- T: What is the equivalent measure in centimeters?
- S: 137 centimeters.
- T: Show the conversion using an equation with an exponent.
- S: 1.37 meters = 1.37 x 10^2 = 137 centimeters.
- T: What is the conversion factor?
- S: 10^2 or 100.

Repeat the sequence with conversion to millimeters, both with multiplication by 10^3 and division by 10^3, 2.6, and 12.08.

Problem 3

A cat weighs 4.5 kilograms. Convert its weight to grams.
A dog weighs 6700 grams. Convert its weight to kilograms.

- T: Work with a partner to find both the cat's weight in grams and the dog's weight in kilograms. Explain your reasoning with an equation using an exponent for each problem.
- S: (Students solve.) 4.5 kg x 10^3 = 4500g and 6700 g ÷ 10^3 = 6.7 kg.
- T: What is the conversion factor for both problems?
- S: 10^3 or 1000.

Repeat this sequence with 2.75 kg to g, and then 6007 g to 6.007 kg and the analogous conversion dividing grams by 10^3 to find the equivalent amount of kilograms.

- T: Let's relate our meter to millimeter measurements to our kilogram to gram conversions.

The most important concept is the equivalence of the two measurements—that is, the weight measurement, like that of the linear measurement, did not change. The change in the type of unit precipitates a change in the number of units. However, the weight has remained the same. Clarify this understanding before moving on to finding the conversion equation by asking, "How can 6007 and 6.007 be equal to each other?" (While the numeric values differ, the unit size is also different. 6007 is grams. 6.007 is kilograms. Kilograms are 1000 times as large as grams. Therefore it takes a lot fewer kilograms to make the same amount as something measured in grams.) Then, lead students to articulate that conversions from largest to smallest units we multiplied by 10^3, to convert from smallest to largest, we need to divide by 10^3.

Lesson 4:	Use exponents to denote powers of 10 with application to metric conversions.
Date:	6/28/13

Problem 4

$0.6 \text{ l} \times 10^3 = 600$ ml; 0.6×10^2; $764 \text{ ml} \div 10^3 = 0.764$ liters

    a. The baker uses 0.6 liter of vegetable oil to bake brownies. How many milliliters of vegetable oil did he use? He is asked to make 100 batches for a customer. How many liters of oil will he need?

    b. After gym class, Mei Ling drank 764 milliliters of water. How many liters of water did she drink?

After solving the baker problem, have students share about what they notice with the measurement conversions thus far.

    S: To convert from kilometers to meters, kilograms to grams, liters to milliliters, we multiplied by a conversion factor of 1000 to get the answer.
→ We multiply with a conversion factor of 100 to convert from meters to centimeters. → When we multiply by 1000, our number shifts 3 spaces to the left on the place value chart. When we divide by 1000, the number shifts 3 spaces to the right. → The smaller the unit is, the bigger the quantity we need to make the same measurement.

Repeat this sequence, converting 1,045 ml to liters and 0.008 l to milliliters. Ask students to make comparisons between and among conversions and conversion factors.

Problem Set (10 minutes)

Students should do their personal best to complete the problem set within the allotted 10 minutes. For some classes, it may be appropriate to modify the assignment by specifying which problems they work on first. Some problems do not specify a method for solving. Students solve these problems using the RDW approach used for Application Problems.

In this Problem Set, we suggest all students begin with Problem 1 and leave Problem 6 to the end if they have time.

Student Debrief (10 minutes)

Lesson Objective: Use exponents to denote powers of 10 and with application to metric conversions.

The Student Debrief is intended to invite reflection and active processing of the total lesson experience.

Invite students to review their solutions for the Problem Set. They should check work by comparing answers with a partner before going over answers as a class. Look for misconceptions or misunderstandings that can be addressed in the Debrief. Guide students in a conversation to debrief the worksheet and process the lesson. You may choose to use any combination of the questions below to lead the discussion.

- Reflect on the kinds of thinking you did on Task 1 and Task 2. How are they alike? How are they different?
- How did you convert centimeters to meters? What is the conversion factor?
- How did you convert meters to centimeters? What is the conversion factor?
- In Task 3, how did you convert from meters to **millimeters**? What conversion factor did you use?
- What can you conclude about the operation you use when converting from a small unit to a large unit? When converting from a large unit to a small unit?
- Students might journal about the meanings of *centi-*, *milli-* and even other units like *deci-* and *deca-*.
- Which is easier for you to think about: converting from larger to smaller units or smaller to larger units? Why? What is the difference in the thinking required to do each?

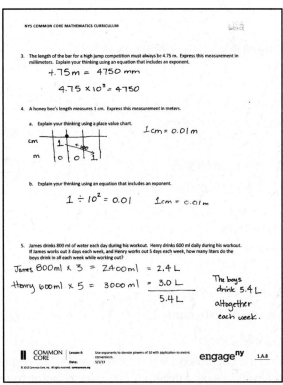

Exit Ticket (3 minutes)

After the Student Debrief, instruct students to complete the Exit Ticket. A review of their work will help you assess the students' understanding of the concepts that were presented in the lesson today and plan more effectively for future lessons. You may read the questions aloud to the students.

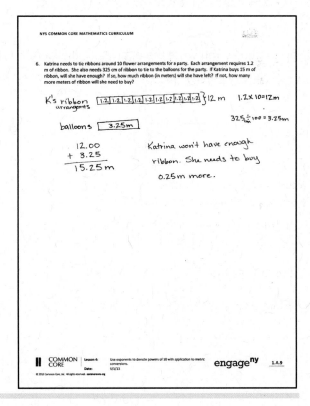

Lesson 4: Use exponents to denote powers of 10 with application to metric conversions.
Date: 6/28/13

COMMON CORE MATHEMATICS CURRICULUM • NY   Lesson 4 Problem Set 5•1

Name _____   Date _____

1. Convert using an equation with an exponent.

   a. 3 meters to centimeters   _____ = _____ cm
   b. 900 centimeters to meters   _____ = _____ m

   c. 8.1 liters to milliliters   _____ = _____ ml
   d. 537 milliliters to liters   _____ = _____ l

   e. 90.5 kilometers to meters   _____ = _____ m
   f. Convert 23 meters to kilometers. _____ = _____ km

   g. 0.4 kilograms to grams   _____ = _____ g
   h. 80 grams to kilograms   _____ = _____ kg

   i. Circle the conversion factor in each equation above. Explain why converting from meters to centimeters uses a different conversion factor than converting from liters to milliliters, kilometers to meters, and kilograms to grams.

2. Read each aloud as you write the equivalent measures.

   a. 3.5 km =   _____ km _____ m
   b. 1.23 l =   _____ l _____ ml
   c. 2.002 kg = _____ kg _____ g
   d. 3 ml =   _____ l
   e. 3012 g =   _____ kg
   f. _____ m =   2.10 cm

Lesson 4: Use exponents to denote powers of 10 with application to metric conversions.
Date: 6/28/13

3. The length of the bar for a high jump competition must always be 4.75 m. Express this measurement in millimeters. Explain your thinking using an equation that includes an exponent.

4. A honey bee's length measures 1 cm. Express this measurement in meters.

   a. Explain your thinking using a place value chart.

   b. Explain your thinking using an equation that includes an exponent.

5. James drinks 800 ml of water each day during his workout. Henry drinks 600 ml daily during his workout. If James works out 3 days each week, and Henry works out 5 days each week, how many liters do the boys drink in all each week while working out?

6. Katrina needs to tie ribbons around 10 flower arrangements for a party. Each arrangement requires 1.2 m of ribbon. She also needs 325 cm of ribbon to tie to the balloons for the party. If Katrina buys 15 m of ribbon, will she have enough? If so, how much ribbon (in meters) will she have left? If not, how many more meters of ribbon will she need to buy?

COMMON CORE MATHEMATICS CURRICULUM • NY   Lesson 4 Exit Ticket 5•1

Name _____   Date _____

1. Convert:

    a. 2 meters to centimeters   2 m × _____ = _____ cm

    b. 40 milliliters to liters   40 ml ÷ _____ = _____ l

2. Read each aloud as you write the equivalent measures.

    a. 4.37 l =   _____ l _____ ml

    b. 81.62 kg =   _____ kg _____ g

Lesson 4: Use exponents to denote powers of 10 with application to metric conversions.
Date: 6/28/13

COMMON CORE MATHEMATICS CURRICULUM • NY   **Lesson 4 Homework** **5•1**

Name _____   Date _____

1. Convert:

   a. 5 meters to centimeters   5 m × _____ = _____ cm

   b. 60 centimeters to meters   60 cm ÷ _____ = _____ m

   c. 2300 milliliters to liters.   2.3 l ÷ _____ = _____ ml

   d. 0.462 liters to milliliters   0.462 l × _____ = _____ ml

   e. 80.4 kilometers to meters   _____ = _____ m

   f. 0.725 kilometers to meters   _____ = _____ m

   g. 456 grams to kilograms   _____ = _____ kg

   h. 0.3 kilograms to grams   _____ = _____ g

2. Read each aloud as you write the equivalent measures.

   a. 2.7 km = _____ km _____ m

   b. 3.46 l = _____ l _____ ml

   c. 5.005 kg = _____ kg _____ g

   d. 8 ml = _____ l

   e. 4079 g = _____ kg

Lesson 4: Use exponents to denote powers of 10 with application to metric conversions.
Date: 6/28/13

3. A dining room table measures 1.78 m long. Express this measurement in millimeters.

   a. Explain your thinking using a place value chart.

   b. Explain your thinking using an equation that includes an exponent.

4. Eric and YiTing commute to school every day. Eric walks 0.81 km and YiTing walks 0.65 km. How far did each of them walk in meters? Explain your answer using an equation that includes an exponent.

5. There were 9 children at a birthday party. Each child drank one 200 ml juice box. How many liters of juice did they drink altogether? Explain your answer using an equation that includes an exponent.

COMMON CORE MATHEMATICS CURRICULUM • NY   Lesson 5   5•1

GRADE 5 • MODULE 1

Topic B
Decimal Fractions and Place Value Patterns

5.NBT.3

Focus Standard:	5.NBT.3	Read, write, and compare decimals to thousandths.
		a. Read and write decimals to thousandths using base-ten numerals, number names, and expanded form, e.g., 347.392 = 3 × 100 + 4 × 10 + 7 × 1 + 3 × (1/10) + 9 × (1/100) + 2 × (1/1000).
		b. Compare two decimals to thousandths based on meanings of the digits in each place, using >, =, and < symbols to record the results of comparisons.
Instructional Days:	2	
Coherence -Links from:	G4–M1	Place Value, Rounding, and Algorithms for Addition and Subtraction
-Links to:	G6–M2	Arithmetic Operations Including Dividing by a Fraction

Naming decimal fractions in expanded, unit, and word forms in order to compare decimal fractions is the focus of Topic B (**5.NBT.3**). Familiar methods of expressing expanded form are used, but students are also encouraged to apply their knowledge of exponents to expanded forms (e.g., 4300.01 = 4 × 10^3 + 3 × 10^2 + 1 × 1/100). Place value charts and disks offer a beginning for comparing decimal fractions to the thousandths, but are quickly supplanted by reasoning about the meaning of the digits in each place and noticing differences in the values of like units and expressing those comparisons with symbols (>, <, and =).

A Teaching Sequence Towards Mastery of Decimal Fractions and Place Value Patterns
Objective 1: Name decimal fractions in expanded, unit, and word forms by applying place value reasoning. (Lesson 5)
Objective 2: Compare decimal fractions to the thousandths using like units and express comparisons with >, <, =. (Lesson 6)

Lesson 5

Objective: Name decimal fractions in expanded, unit, and word forms by applying place value reasoning.

Suggested Lesson Structure

■ Fluency Practice   (12 minutes)
■ Application Problems (8 minutes)
■ Concept Development   (30 minutes)
■ Student Debrief   (10 minutes)
  Total Time   **(60 minutes)**

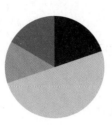

Fluency Practice (12 minutes)

- Multiply Decimals by 10, 100, and 1000   **5.NBT.2**   (8 minutes)
- Multiply and Divide by Exponents   **5.NBT.2**   (2 minutes)
- Multiply Metric Units   **5.MD.1**   (2 minutes)

Sprint: Multiply Decimals by 10, 100, and 1000 (8 minutes)

Materials: (S) Multiply Decimals by 10, 100, and 1000 Sprint

Note: This Sprint will help students work towards automaticity of multiplying and dividing decimals by 10, 100, and 1000.

Multiply and Divide by Exponents (2 minutes)

Materials: (S) Personal white boards

Note: This fluency will help students work towards mastery on the concept that was introduced in Lesson 4.

Depending on students' depth of knowledge, this fluency may be done with support from a personal place value chart or done simply by responding on the personal white board with the product or quotient.

T: (Project place value chart from millions to thousandths.) Write 54 tenths as a decimal.
S: (Students write 5 in the ones column and 4 in the tenths column.)
T: Say the decimal.
S: 5.4
T: Multiply it by 10^2.
S: (Students indicate change in value by using arrows from each original place value to product or

Lesson 5:	Name decimal fractions in expanded, unit, and word forms by applying place value reasoning.
Date:	6/28/13

© 2013 Common Core, Inc. All rights reserved. commoncore.org

COMMON CORE MATHEMATICS CURRICULUM • NY   Lesson 5   5•1

quotient on personal white board. They may, instead, simply write product.)
T: Say the product.
S: 540.

Repeat the process and sequence for 0.6×10^2, $0.6 \div 10^2$, 2.784×10^3, and $6583 \div 10^3$.

Multiplying Metric Units (2 minutes)

Materials: (S) Personal white boards

Note: This fluency will help students work towards mastery on the concept that was introduced in Lesson 4.

T: (Write 3 m = ___ cm.) Show 3 in your place value chart.
S: (Students write 3 in the ones column.)
T: How many centimeters are in 1 meter?
S: 100 centimeters.
T: Show how many centimeters are in 3 meters on your place value chart.
S: (Students cross out the 3 and shift it 2 place values to the left to show 300.)
T: How many centimeters are in 3 meters?
S: 300 centimeters.

Repeat the process and procedure for 7 kg = ____ g, 7000 ml = ____ l, 7500 m = ____ km ____ m, and 8350 g = ____ kg ____ g.

Application Problems (8 minutes)

Jordan measures a desk at 200 cm. James measures the same desk in millimeters, and Amy measures the same desk in meters. What is James measurement in millimeters? What is Amy's measurement in meters? Show your thinking using a place value mat or equation using place value mat or an equation with exponents.

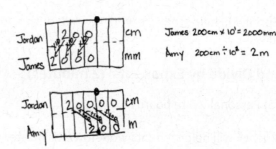

Note: Today's application problem offers students a quick review of yesterday's concepts before moving forward to naming decimals.

Concept Development (30 minutes)

Materials: (S) Personal white board with place value chart

Opener

T: (Write *three thousand forty seven* on the board.) On your personal white board, write this number

Lesson 5: Name decimal fractions in expanded, unit, and word forms by applying place value reasoning.
Date: 6/28/13

1.B.3

in standard form, expanded form, and unit form.

- T: Explain to your partner the purpose of writing this number in these different forms.
- S: Standard form shows us the digits that we are using to represent that amount. → Expanded form shows how much each digit is worth and that the number is a total of those values added together. → Unit form helps us see how many of each size unit are in the number.

Problem 1

Represent 1 thousandth and 3 thousandths in standard, expanded, and unit form.

- T: Write one thousandth using digits on your place value chart.
- T: How many ones, tenths, hundredths, thousandths?
- S: Zero, zero, zero, one.
- T: This is the standard form of the decimal for 1 thousandth.
- T: We write 1 thousandth as a fraction like this. (Write $\frac{1}{1000}$ on the board.)
- T: 1 thousandth is a single copy of a thousandth. I can write the expanded form using a fraction like this, $1 \times \left(\frac{1}{1000}\right)$ (saying one copy of one thousandth) or using a decimal like this 1×0.001. (Write on the board.)
- T: The unit form of this decimal looks like this *1 thousandth*. (Write on the board.) We use a numeral (point to 1) and the unit (point to thousandth) written as a word.

MP.7

> One thousandth = $0.001 = \frac{1}{1000}$
>
> $\frac{1}{1000} = 1 \times \left(\frac{1}{1000}\right)$
>
> $0.001 = 1 \times 0.001$
>
> 1 thousandth

- T: Imagine 3 copies of 1 thousandth. How many thousandths is that?
- S: 3 thousandths.
- T: (Write in standard form and as a fraction.)
- T: 3 thousandths is 3 copies of 1 thousandth, turn and talk to your partner about how this would be written in expanded form using a fraction and using a decimal.

> Three thousandths = $0.003 = \frac{3}{1000}$
>
> $\frac{3}{1000} = 3 \times \left(\frac{1}{1000}\right)$
>
> $0.003 = 3 \times 0.001$
>
> 3 thousandths

Lesson 5:	Name decimal fractions in expanded, unit, and word forms by applying place value reasoning.
Date:	6/28/13

1.B.4

COMMON CORE MATHEMATICS CURRICULUM • NY   Lesson 5   5•1

Problem 2

Represent 13 thousandths in standard, expanded, and unit form.

T: Write thirteen thousandths in standard form, and expanded form using fractions and then using decimals. Turn and share with your partner.

S: Zero point zero one three is standard form. Expanded forms are

$1 \times \left(\frac{1}{100}\right) + 3 \times \left(\frac{1}{1000}\right)$ and $1 \times 0.01 + 3 \times 0.001$.

T: Now write this decimal in unit form.

S: 1 hundredth 3 thousandths → 13 thousandths.

T: (Circulate and write responses on the board.) I notice that there seems to be more than one way to write this decimal in unit form. Why?

S: This is 13 copies of 1 thousandth. → You can write the units separately or write the 1 hundredth as 10 thousandths. You add 10 thousandths and 3 thousandths to get 13 thousandths.

NOTES ON MULTIPLE MEANS FOR ENGAGEMENT:

Students struggling with naming decimals using different unit forms may benefit from a return to concrete materials. Using place value disks to make trades for smaller units combined with place value understandings from Lessons 1 and 2 help make the connection between 1 hundredth 3 thousandths and 13 thousandths.

It may also be fruitful to invite students to extend their Grade 4 experiences with finding equivalent fractions for tenths and hundredths to finding equivalent fraction representations in thousandths.

```
Thirteen thousandths = 0.013 = 13/1000
                       13/1000 =
0.013 = 1 × 0.01 + 3 × 0.001
1 hundredth 3 thousandths
13 thousandths
```

Repeat with 0.273 and 1.608 allowing students to combine units in their unit forms (for example, 2 tenths 73 thousandths; 273 thousandths; 27 hundredths 3 thousandths). Use more or fewer examples as needed reminding students who need it that *and* indicates the decimal in word form.

Problem 3

Represent 25.413 in word, expanded, and unit form.

T: (Write on the board.) Write 25.413 in word form on your board. (Students write.)

S: Twenty-five and four hundred thirteen thousandths.

T: Now, write this decimal in unit form on your board.

S: 2 tens 5 ones 4 tenths 1 hundredth 3 thousandths.

T: What are other unit forms of this number?

Allow students to combine units, e.g., 25 ones 413 thousandths, 254 tenths 13 hundredths, 25,413 thousandths.

Lesson 5: Name decimal fractions in expanded, unit, and word forms by applying place value reasoning.
Date: 6/28/13

1.B.5

COMMON CORE MATHEMATICS CURRICULUM • NY   Lesson 5   5•1

T: Write it as a mixed number, then in expanded form. Compare your work with your partner's.

> Twenty-five and four hundred thirteen thousandths = $25\frac{413}{1000}$ = 25.413
>
> $25\frac{413}{1000} = 2 \times 10 + 5 \times 1 + 4 \times \left(\frac{1}{10}\right) + 1 \times \left(\frac{1}{100}\right) + 3 \times \left(\frac{1}{1000}\right)$
>
> $25.413 = 2 \times 10 + 5 \times 1 + 4 \times 0.1 + 1 \times 0.01 + 3 \times 0.001$
>
> 2 tens 5 ones 4 tenths 1 hundredths 3 thousandths
>
> 25 ones 413 thousandths

Repeat the sequence with 12.04 and 9.495. Use more or fewer examples as needed.

Problem 4

Write the standard, expanded, and unit forms of *four hundred four thousandths* and *four hundred and four thousandths*.

T: Work with your partner to write these decimals in standard form. (Circulate looking for misconceptions about the use of the word *and*.)

T: Tell the digits you used to write *four hundred four thousandths*.

T: How did you know where to write the decimal in the standard form?

S: The word *and* tells us where the fraction part of the number starts.

T: Now work with your partner to write the expanded and unit forms for these numbers.

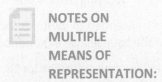

NOTES ON MULTIPLE MEANS OF REPRESENTATION:

Guide students to draw on their past experiences with whole numbers and make parallels to decimals. Whole number units are named by smallest base thousand unit, e.g., 365,000 = 365 thousand and 365 = 365 ones. Likewise, we can name decimals by the smallest unit (e.g., 0.63 = 63 hundredths).

> Four hundred four thousandths = $\frac{404}{1000}$ = 0.404
>
> $\frac{404}{1000} = 4 \times \left(\frac{1}{10}\right) + 4 \times \left(\frac{1}{1000}\right)$
>
> $0.404 = 4 \times 0.1 + 4 \times 0.001$
>
> 4 tenths 4 thousandths
>
> Four hundred and four thousandths = $400\frac{4}{1000}$ =
> 400.004 $400\frac{4}{1000} = 4 \times 100 + 4 \times \left(\frac{1}{1000}\right)$

Repeat this sequence with *two hundred two thousandths* and *nine hundred and nine tenths*.

T: Work on your problem set now. Read the word forms carefully!

Lesson 5: Name decimal fractions in expanded, unit, and word forms by applying place value reasoning.
Date: 6/28/13

1.B.6

Problem Set (10 minutes)

Students should do their personal best to complete the problem set within the allotted 10 minutes. For some classes, it may be appropriate to modify the assignment by specifying which problems they work on first. Some problems do not specify a method for solving. Students solve these problems using the RDW approach used for Application Problems.

Student Debrief (10 minutes)

Lesson Objective: Name decimal fractions in expanded, unit, and word forms by applying place value reasoning.

The Student Debrief is intended to invite reflection and active processing of the total lesson experience.

Invite students to review their solutions for the problem set. They should check work by comparing answers with a partner before going over answers as a class. Look for misconceptions or misunderstandings that can be addressed in the Debrief. Guide students in a conversation to debrief the Problem Set and process the lesson. You may choose to use any combination of the questions below to lead the discussion.

- Which tasks in Problem 1 are alike? Why?
- What is the purpose of writing a decimal number in expanded form using fractions? What was the objective of our lesson today?
- Compare your answers to Problem 1(c) and 1(d). What is the importance of the word *and* when naming decimals in standard form?
- When might expanded form be useful as a calculation tool? (It helps us see the like units, could help to add and subtract mentally.)
- How is expanded form related to the standard form of a number?

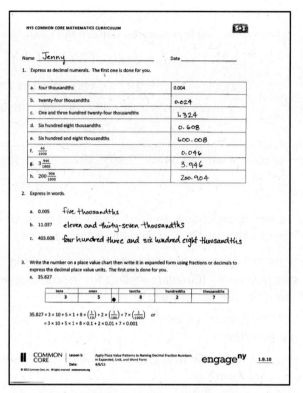

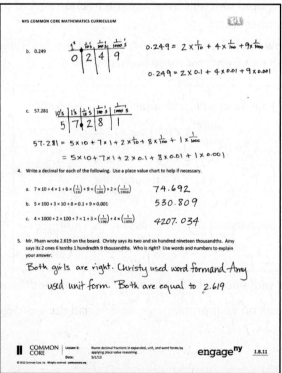

Exit Ticket (3 minutes)

After the Student Debrief, instruct students to complete the Exit Ticket. A review of their work will help you assess the students' understanding of the concepts that were presented in the lesson today and plan more effectively for future lessons. You may read the questions aloud to the students.

Lesson 5:	Name decimal fractions in expanded, unit, and word forms by applying place value reasoning.
Date:	6/28/13

A

Multiply.   # Correct _____

#	Problem	Answer	#	Problem	Answer
1	62.3 x 10 =		23	4.1 x 1000 =	
2	62.3 x 100 =		24	7.6 x 1000 =	
3	62.3 x 1000 =		25	0.01 x 1000 =	
4	73.6 x 10 =		26	0.07 x 1000 =	
5	73.6 x 100 =		27	0.072 x 100 =	
6	73.6 x 1000 =		28	0.802 x 10 =	
7	0.6 x 10 =		29	0.019 x 1000 =	
8	0.06 x 10 =		30	7.412 x 1000 =	
9	0.006 x 10 =		31	6.8 x 100 =	
10	0.3 x 10 =		32	4.901 x 10 =	
11	0.3 x 100 =		33	16.07 x 100 =	
12	0.3 x 1000 =		34	9.19 x 10 =	
13	0.02 x 10 =		35	18.2 x 100 =	
14	0.02 x 100 =		36	14.7 x 1000 =	
15	0.02 x 1000 =		37	2.021 x 100 =	
16	0.008 x 10 =		38	172.1 x 10 =	
17	0.008 x 100 =		39	3.2 x 20 =	
18	0.008 x 1000 =		40	4.1 x 20 =	
19	0.32 x 10 =		41	3.2 x 30 =	
20	0.67 x 10 =		42	1.3 x 30 =	
21	0.91 x 100 =		43	3.12 x 40 =	
22	0.74 x 100 =		44	14.12 x 40 =	

© Bill Davidson

Lesson 5: Name decimal fractions in expanded, unit, and word forms by applying place value reasoning.
Date: 6/28/13

B Improvement _____ # Correct _____

Multiply.

#	Problem		#	Problem	
1	46.1 x 10 =		23	5.2 x 1000 =	
2	46.1 x 100 =		24	8.7 x 1000 =	
3	46.1 x 1000 =		25	0.01 x 1000 =	
4	89.2 x 10 =		26	0.08 x 1000 =	
5	89.2 x 100 =		27	0.083 x 10 =	
6	89.2 x 1000 =		28	0.903 x 10 =	
7	0.3 x 10 =		29	0.017 x 1000 =	
8	0.03 x 10 =		30	8.523 x 1000 =	
9	0.003 x 10 =		31	7.9 x 100 =	
10	0.9 x 10 =		32	5.802 x 10 =	
11	0.9 x 100 =		33	27.08 x 100 =	
12	0.9 x 1000 =		34	8.18 x 10 =	
13	0.04 x 10 =		35	29.3 x 100 =	
14	0.04 x 100 =		36	25.8 x 1000 =	
15	0.04 x 1000 =		37	3.032 x 100 =	
16	0.007 x 10 =		38	283.1 x 10 =	
17	0.007 x 100 =		39	2.1 x 20 =	
18	0.007 x 1000 =		40	3.3 x 20 =	
19	0.45 x 10 =		41	3.1 x 30 =	
20	0.78 x 10 =		42	1.2 x 30 =	
21	0.28 x 100 =		43	2.11 x 40 =	
22	0.19 x 100 =		44	13.11 x 40 =	

© Bill Davidson

Lesson 5: Name decimal fractions in expanded, unit, and word forms by applying place value reasoning.
Date: 6/28/13

Name _____ **Date** _____

1. Express as decimal numerals. The first one is done for you.

a. four thousandths	0.004
b. twenty-four thousandths	
c. one and three hundred twenty-four thousandths	
d. six hundred eight thousandths	
e. six hundred and eight thousandths	
f. $\frac{46}{1000}$	
g. $3\frac{946}{1000}$	
h. $200\frac{904}{1000}$	

2. Express in words.

   a. 0.005

   b. 11.037

   c. 403.608

3. Write the number on a place value chart then write it in expanded form using fractions or decimals to express the decimal place value units. The first one is done for you.

   a. 35.827

tens	ones		tenths	hundredths	thousandths
3	5	●	8	2	7

$35.827 = 3 \times 10 + 5 \times 1 + 8 \times \left(\frac{1}{10}\right) + 2 \times \left(\frac{1}{100}\right) + 7 \times \left(\frac{1}{1000}\right)$ or

$= 3 \times 10 + 5 \times 1 + 8 \times 0.1 + 2 \times 0.01 + 7 \times 0.001$

Lesson 5: Name decimal fractions in expanded, unit, and word forms by applying place value reasoning.
Date: 6/28/13

b. 0.249

c. 57.281

4. Write a decimal for each of the following. Use a place value chart to help if necessary.

   a. $7 \times 10 + 4 \times 1 + 6 \times \left(\frac{1}{10}\right) + 9 \times \left(\frac{1}{100}\right) + 2 \times \left(\frac{1}{1000}\right)$

   b. $5 \times 100 + 3 \times 10 + 8 \times 0.1 + 9 \times 0.001$

   c. $4 \times 1000 + 2 \times 100 + 7 \times 1 + 3 \times \left(\frac{1}{100}\right) + 4 \times \left(\frac{1}{1000}\right)$

5. Mr. Pham wrote 2.619 on the board. Christy says its two and six hundred nineteen thousandths. Amy says its 2 ones 6 tenths 1 hundredth 9 thousandths. Who is right? Use words and numbers to explain your answer.

Name _____ Date _____

1. Express nine thousandths as a decimal.

2. Express twenty-nine thousandths as a fraction.

3. Express 24.357 in words.

    a. Write the expanded form using fractions or decimals.

    b. Express in unit form.

COMMON CORE MATHEMATICS CURRICULUM • NY   Lesson 5 Homework   5•1

Name _____   Date _____

1. Express as decimal numerals. The first one is done for you.

a. Five thousandths	0.005
b. Thirty-five thousandths	
c. Nine and two hundred thirty-five thousandths	
d. Eight hundred and five thousandths	
e. $\frac{8}{1000}$	
f. $\frac{28}{1000}$	
g. $7\frac{528}{1000}$	
h. $300\frac{502}{1000}$	

2. Express in words.

   a. 0.008

   b. 15.062

   c. 607.409

3. Write the number on a place value chart then write it in expanded form using fractions or decimals to express the decimal place value units. The first one is done for you.

   a. 27.346

tens	ones	•	tenths	hundredths	thousandths
2	7	•	3	4	6

   $27.346 = 2 \times 10 + 7 \times 1 + 3 \times \left(\frac{1}{10}\right) + 4 \times \left(\frac{1}{100}\right) + 6 \times \left(\frac{1}{1000}\right)$

   OR

   $27.346 = 2 \times 10 + 7 \times 1 + 3 \times 0.1 + 4 \times 0.01 + 6 \times 0.001$

Lesson 5: Name decimal fractions in expanded, unit, and word forms by applying place value reasoning.
Date: 6/28/13

1.B.14

b. 0.362

c. 49.564

4. Write a decimal for each of the following. Use a place value chart to help if necessary.

   a. $3 \times 10 + 5 \times 1 + 2 \times \left(\frac{1}{10}\right) + 7 \times \left(\frac{1}{100}\right) + 6 \times \left(\frac{1}{1000}\right)$

   b. $9 \times 100 + 2 \times 10 + 3 \times 0.1 + 7 \times 0.001$

   c. $5 \times 1000 + 4 \times 100 + 8 \times 1 + 6 \times \left(\frac{1}{100}\right) + 5 \times \left(\frac{1}{1000}\right)$

5. At the beginning of a lesson, a piece of chalk is 2.967 of an inch. At the end of lesson, it's 2.308 of an inch. Write the two amounts in expanded form using fractions.

   a. At the beginning of the lesson:

   b. At the end of the lesson:

6. Mrs. Herman asked the class to write an expanded form for 412.638. Nancy wrote the expanded form using fractions and Charles wrote the expanded form using decimals. Write their responses.

COMMON CORE MATHEMATICS CURRICULUM • NY   Lesson 6   5•1

Lesson 6

Objective: Compare decimal fractions to the thousandths using like units and express comparisons with >, <, =.

Suggested Lesson Structure

- Fluency Practice   (12 minutes)
- Application Problems   (8 minutes)
- Concept Development   (30 minutes)
- Student Debrief   (10 minutes)

  Total Time   **(60 minutes)**

Fluency Practice (12 minutes)

- Find the Midpoint **5.NBT.4**   (5 minutes)
- Rename the Units **5.NBT.1**   (2 minutes)
- Multiply by Decimal Fractions **5.NBT.3a** (5 minutes)

Find the Midpoint (5 minutes)

Materials: (S) Personal white boards

Note: Practicing this skill in isolation will help students conceptually understand rounding decimals in lesson 12.

T: (Project a 0 on the left side of a number line and 10 on the right side of the number line.) What's halfway between 0 ones and 10 ones?

S: 5 ones.

T: (Write 5 ones halfway between the 0 and 10. Draw a second number line directly beneath the first. Write 0 on the left side and 1 on the right side.) How many tenths is 1?

S: 1 is 10 tenths.

T: (Write 10 tenths below the 1.) On your boards, write the decimal that is halfway between 0 and 1 or 10 tenths?

S: (Students write 0.5 approximately halfway between 0 and 1 on their number lines.)

Repeat the process for these possible sequences: 0 and 0.1; 0 and 0.01; 10 and 20; 1 and 2; 0.1 and 0.2; 0.01 and 0.02; 0.7 and 0.8; 0.7 and 0.71; 9 and 10; 0.9 and 1; and 0.09 and 0.1.

Lesson 6:	Compare decimal fractions to the thousandths using like units and express comparisons with >, <, and =.
Date:	6/28/13

1.B.16

COMMON CORE MATHEMATICS CURRICULUM • NY   Lesson 6   5•1

Rename the Units (2 minutes)

Note: Reviewing unit conversions will help students work towards mastery of decomposing common units into compound units.

    T: (Write 100 cm = ____ m.) Rename the units.
    S: 100 cm = 1 meter.
    T: (Write 200 cm = ____ m.) Rename the units.
    S: 200 centimeters = 2 meters.
    T: 700 centimeters.
    S: 7 meters.
    T: (Write 750 cm = ____ m ____ cm.) Rename the units.
    S: 7 meters 50 centimeters.

Repeat the process for 450 cm, 630 cm, and 925 cm.

Multiply by Decimal Fractions (5 minutes)

Materials: (S) Personal white boards, place value charts to the thousandths

Notes: Review will help students work towards mastery of this skill, which was introduced in previous lessons.

    T: (Project a place value chart from tens to thousandths. Beneath the chart, write 3 x 10 = ____ .) Say the multiplication sentence.
    S: 3 x 10 = 30.
    T: (Write 3 in the tens column. Below the multiplication sentence write 30. To the right of 3 x 10, write 4 x 1 = ____ .) Say the multiplication sentence.
    S: 4 x 1 = 4.
    T: (Write 4 in the ones column and fill in the addition sentence so that it reads 30 + 4.)

Repeat process with each of the equations below so that in the end, the number 34.652 will be written in the place value chart and 30 + 4 + 0.6 + 0.05 + 0.002 is written underneath it:

$$6 \times \frac{1}{10} \qquad 5 \times \frac{1}{100} \qquad 2 \times \frac{1}{1000}$$

    T: Say the addition sentence.
    S: 30 + 4 + 0.6 + 0.05 + 0.002 = 34.652.
    T: (Write 75.614 on the place value chart.) Write the number in expanded form.

Repeat for these possible sequences: 75.604; 20.197; and 40.803.

Lesson 6: Compare decimal fractions to the thousandths using like units and express comparisons with >, <, and =.
Date: 6/28/13

COMMON CORE MATHEMATICS CURRICULUM • NY

Lesson 6 5•1

Application Problems (8 minutes)

Ms. Meyer measured the edge of her dining table to the thousandths of a meter. The edge of the table measured 32.15 meters. Write her measurement in word form, unit form, and in expanded form using fractions and decimals.

(Encourage students to name the decimal by decomposing it into various units, e.g., 3,215 hundredths; 321 tenths 5 hundredths; 32 ones 15 hundredths.)

Concept Development (30 minutes)

Materials: (S) Place value chart and marker

Problem 1

Compare 13,196 and 13,296.

MP.2

- T: (Point to 13,196.) Read the number.
- S: (Students read number.)
- T: (Point to 13,296.) Read the number.
- S: (Students read number.)
- T: Which number is larger? How can you tell?
- S: 13,296 is larger than 13,196 because the digit in the hundreds place is one bigger. → 13,296 is 100 more than 13,196. → 13,196 has 131 hundreds and 13,296 has 132 hundreds, so 13,296 is greater.
- T: Use a symbol to show which number is greater.
- S: 13,196 < 13,296

Problem 2

Compare 0.012 and 0.002.

- T: Write 2 thousandths in standard form on your place value chart. (Write 2 thousandths on the board.)
- S: (Students write.)
- T: Say the digits that you wrote on your chart.
- S: Zero point zero zero two.
- T: Write 12 thousandths in standard form underneath 0.002 on your chart. (Write 12 thousandths on the board.)
- S: (Students write.)
- T: Say the digits that you wrote on your chart.
- S: Zero point zero one two.
- T: Say this number in unit form.
- S: 1 hundredth 2 thousandths.

Lesson 6:	Compare decimal fractions to the thousandths using like units and express comparisons with >, <, and =.
Date:	6/28/13

1.B.18

T: Which number is larger? Turn and talk to your partner about how you can decide.

S: 0.012 is bigger than 0.002. → In 0.012, there is a one in the hundredths place, but 0.002 has a zero in the hundredths so that means 0.012 is bigger than 0.002. → 12 of something is greater than 2 of the same thing. Just like 12 apples are more than 2 apples.

Next, you might have the students write the two numbers on the place value chart and move from largest units to smallest. Close by writing 0.002 < 0.012.

Problem 3

Compare $\frac{299}{1000}$ and $\frac{3}{10}$.

T: Write 3 tenths in standard form on your place value chart.

S: (Students write.)

T: Write 299 thousandths in standard form on your place value chart under 3 tenths.

S: (Students write.)

T: Which decimal has more tenths?

S: 0.3

T: If we traded 3 tenths for thousandths, how many thousandths would we need? Turn and talk to your partner.

S: 300 thousandths.

T: Name these decimals using unit form and compare. Tell your partner which is more.

S: 299 thousandths; 300 thousandths is more.

T: Show this relationship with a symbol.

S: 0.299 < 0.3

Repeat the sequence with $\frac{705}{1000}$ and $\frac{7}{10}$ and 15.203 and 15.21.

Encourage students to name the fractions and decimals using like units as above, e.g., 15 ones 20 tenths 3 hundredths and 15 ones 21 tenths 0 hundredths or 15,203 thousandths and 15,210 thousandths. Be sure to have students express the relationships using <, >, and =.

Problem 4

Order from least to greatest: 0.413, 0.056, 0.164, and 0.531.

Have students order the decimals then explain their strategies (unit form, using place value chart to compare largest to smallest unit looking for differences in value).

NOTES ON MULTIPLE MEANS OF ENGAGEMENT:

Help students deepen their understanding of comparing decimals by returning to concrete materials. Some students may not see that 0.4 > 0.399 because they are focusing on the number of digits to the right of the decimal rather than their value. Comparison of like units becomes a concrete experience when students' attention is directed to comparisons of largest to smallest place value on the chart and when they are encouraged to make trades to the smaller unit using disks.

NOTES ON MULTIPLE MEANS OF ENGAGEMENT:

Provide an extension by including fractions along with decimals to be ordered.

Order from least to greatest: 29.5, 27.019, and $27\frac{5}{1000}$.

Repeat with the following in ascending and descending order: 27.005; 29.04; 27.019; 29.5; 119.177; 119.173; 119.078; and 119.18.

Problem Set (10 minutes)

Students should do their personal best to complete the problem set within the allotted 10 minutes. For some classes, it may be appropriate to modify the assignment by specifying which problems they work on first. Some problems do not specify a method for solving. Students solve these problems using the RDW approach used for Application Problems.

On this problem set, we suggest all students begin with Problems 1, 2, and 5 and possibly leave Problems 3 and 6 to the end if they still have time.

Student Debrief (10 minutes)

Lesson Objective: Compare decimal fractions to the thousandths using like units and express comparisons with >, <, =.

The Student Debrief is intended to invite reflection and active processing of the total lesson experience.

Invite students to review their solutions for the problem set. They should check work by comparing answers with a partner before going over answers as a class. Look for misconceptions or misunderstandings that can be addressed in the Debrief. Guide students in a conversation to debrief the problem set and process the lesson. You may choose to use any combination of the questions below to lead the discussion.

- How is comparing whole numbers like comparing decimal fractions? How is it different?
- You learned two strategies to help you compare numbers (finding a common unit and looking at the place value chart). Which strategy do you like best? Explain.
- Allow sufficient time for in depth discussion of Problem 5. As these are commonly held misconceptions when comparing decimals, it is

worthy of special attention. Ask: What is the mistake that Lance is making? (He's not using like units to compare the numbers. He's forgetting that decimals are named by their smallest units.) How could Angel have named his quantity of water so that the units were the same as Lance's? How would using the same units have helped Lance to make a correct comparison? How is renaming these decimals in the same unit like changing fractions to like denominators?

- Compare 7 tens and 7 tenths. How are they alike? How are they different? (Encourage students to notice that both quantities are 7, but units have different values.) Also, encourage students to notice that they are placed symmetrically in relation to the ones place on place value chart. Tens are 10 times as large as ones while tenths are 1/10 as much. You can repeat with other values, (e.g., 2000, 0.002) or ask students to generate values which are symmetrically placed on the chart.

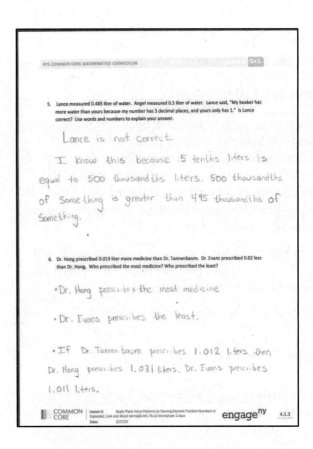

Exit Ticket (3 minutes)

After the Student Debrief, instruct students to complete the Exit Ticket. A review of their work will help you assess the students' understanding of the concepts that were presented in the lesson today and plan more effectively for future lessons. You may read the questions aloud to the students.

Lesson 6: Compare decimal fractions to the thousandths using like units and express comparisons with >, <, and =.
Date: 6/28/13

Lesson 6 Problem Set 5•1

Name _____ Date _____

1. Show the numbers on the place value chart using digits. Use >, <, or = to compare. Explain your thinking to the right.

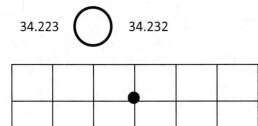

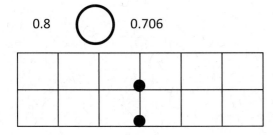

2. Use >, <, or = to compare the following. Use a place value chart to help if necessary.

a.	16.3	◯	16.4
b.	0.83	◯	$\frac{83}{100}$
c.	$\frac{205}{1000}$	◯	0.205
d.	95.580	◯	95.58
e.	9.1	◯	9.099
f.	8.3	◯	83 tenths
g.	5.8	◯	Fifty-eight hundredths

Lesson 6: Compare decimal fractions to the thousandths using like units and express comparisons with >, <, and =.
Date: 6/28/13

h. Thirty-six and nine thousandths	◯	4 tens
i. 202 hundredths	◯	2 hundreds and 2 thousandths
j. One hundred fifty-eight thousandths	◯	158,000
k. 4.15	◯	415 tenths

3. Arrange the numbers in increasing order.

   a. 3.049   3.059   3.05   3.04

   _____

   b. 182.205   182.05   182.105   182.025

   _____

4. Arrange the numbers in decreasing order.

   a. 7.608   7.68   7.6   7.068

   _____

   b. 439.216   439.126   439.612   439.261

   _____

5. Lance measured 0.485 liter of water. Angel measured 0.5 liter of water. Lance said, "My beaker has more water than yours because my number has 3 decimal places and yours only has 1." Is Lance correct? Use words and numbers to explain your answer.

6. Dr. Hong prescribed 0.019 liter more medicine than Dr. Tannenbaum. Dr. Evans prescribed 0.02 less than Dr. Hong. Who prescribed the most medicine? Who prescribed the least? Explain how you know using a place value chart.

Name _____ Date _____

1. Show the numbers on the place value chart using digits. Use >, <, or = to compare. Explain your thinking to the right.

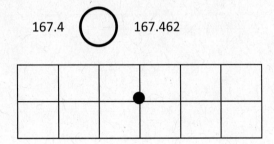

2. Use >, <, and = to compare the numbers.

3. Arrange in descending order.

   76.342   76.332   76.232   76.343

COMMON CORE MATHEMATICS CURRICULUM • NY   Lesson 6 Homework   5•1

Name _____   Date _____

1. Use >, <, or = to compare the following.

a. 16.45	◯	16.454
b. 0.83	◯	$\frac{83}{100}$
c. $\frac{205}{1000}$	◯	0.205
d. 95.045	◯	95.545
e. 419.10	◯	419.099
f. Five ones and eight tenths	◯	Fifty-eight tenths
g. Thirty-six and nine thousandths	◯	Four tens
h. One hundred four and twelve hundredths	◯	One hundred four and two thousandths
i. One hundred fifty-eight thousandths	◯	0.58
j. 703.005	◯	Seven hundred three and five hundredths

2. Arrange the numbers in increasing order.

   a. 8.08   8.081   8.09   8.008

   _____

   b. 14.204   14.200   14.240   14.210

   _____

Lesson 6: Compare decimal fractions to the thousandths using like units and express comparisons with >, <, and =.
Date: 6/28/13

3. Arrange the numbers in decreasing order.

   a. 8.508   8.58   7.5   7.058

   _____

   b. 439.216   439.126   439.612   439.261

   _____

4. James measured his hand. It was 0.17 meters. Jennifer measured her hand. It was 0.165 meters. Whose hand is bigger? How do you know?

5. In a paper airplane contest, Marcel's plane travels 3.345 meters. Salvador's plane travels 3.35 meters. Jennifer's plane travels 3.3 meters. Based on the measurements, whose plane traveled the farthest distance? Whose plane traveled the shortest distance? Explain your reasoning using a place value chart.

COMMON CORE MATHEMATICS CURRICULUM • NY

Lesson 7 5•1

GRADE 5 • MODULE 1

Topic C
Place Value and Rounding Decimal Fractions

5.NBT.4

Focus Standard:	5.NBT.4	Use place value understanding to round decimals to any place.
Instructional Days:	2	
Coherence -Links from:	G4–M1	Place Value, Rounding, and Algorithms for Addition and Subtraction
-Links to:	G6–M2	Arithmetic Operations Including Dividing by a Fraction

Students generalize their knowledge of rounding whole numbers to round decimal numbers to any place in Topic C. In Grades 2 and 4, vertical number lines provided a platform for students to round whole numbers to any place. In Grade 5, vertical number lines again provide support for students to make use of patterns in the base ten system allowing knowledge of whole number rounding (**4.NBT.3**) to be easily applied to rounding decimal values (**5.NBT.4**). The vertical number line is used initially to find more than or less than half way between multiples of decimal units. In Lesson 8, students are encouraged to reason more abstractly as they use place value understanding to approximate by using nearest multiples. Naming those nearest multiples is an application of flexibly naming decimals using like place value units. To round 53.805 (53 ones 805 thousandths) to the nearest hundredth, students find the nearest multiples, 53.800 (53 ones 800 thousandths) and 53.810 (53 ones 810 thousandths) and then decide that 53.805 is exactly halfway between and, therefore, must be rounded up to 53.810.

A Teaching Sequence Towards Mastery of Place Value and Rounding Decimal Fractions
Objective 1: Round a given decimal to any place using place value understanding and the vertical number line. (Lessons 7–8)

Topic C: Place Value and Rounding Decimal Fractions
Date: 6/28/13

1.C.1

© 2013 Common Core, Inc. All rights reserved. commoncore.org

Lesson 7

Objective: Round a given decimal to any place using place value understanding and the vertical number line.

Suggested Lesson Structure

- ■ Fluency Practice   (12 minutes)
- ■ Application Problems   (8 minutes)
- ■ Concept Development   (30 minutes)
- ■ Student Debrief   (10 minutes)
- **Total Time**   **(60 minutes)**

NOTES ON MULTIPLE MEANS OF REPRESENTATION:

Vertical number lines may be a novel representation for students. Their use offers an important scaffold for students' understanding of rounding in that numbers are quite literally rounded up and down to the nearest multiple rather than left or right as in a horizontal number line. Consider showing both a horizontal and vertical line and comparing their features so that students can see the parallels and gain comfort in the use of the vertical line.

Fluency Practice (12 minutes)

- Find the Midpoint **5.NBT.4**   (7 minutes)
- Compare Decimal Fractions **5.NBT.3b** (2 minutes)
- Rename the Units **5.NBT.2**   (3 minutes)

Sprint: Find the Midpoint (7 minutes)

Materials: (S) Personal white boards

Note: Practicing this skill in isolation will help students conceptually understand rounding decimals in this lesson.

Compare Decimal Fractions (2 minutes)

Materials: (S) Personal white boards

Note: This review fluency will help students work towards mastery of comparing decimal numbers, a topic they were introduced to in Lesson 6.

T: (Write 12.57 ___ 12.75.) On your personal boards, compare the numbers using the greater than, less than, or equal sign.

S: (Write 12.57 < 12.75 on boards.)

Repeat the process and procedure:

$0.67 \_\_ \frac{67}{100}$ $\frac{83}{100} \_\_ 0.084$ $328.2 \_\_ 328.099$

NOTES ON MULTIPLE MEANS OF ENGAGEMENT:

Sprints like Compare Fractions may be made more active by allowing students to stand and use their arms to make the >, <, and = signs in response to teacher's question on board.

COMMON CORE MATHEMATICS CURRICULUM • NY   Lesson 7   5•1

 4.07 __ forty-seven tenths   twenty-four and 9 thousandths ___ 3 tens

Rename the Units (3 minutes)

Note: Renaming decimals using various units strengthens student understanding of place value and provides an anticipatory set for rounding decimals in Lessons 7 and 8.

- T: (Write 1.5 = ____ tenths.) Fill in the blank.
- S: 15 tenths.
- T: (Write 1.5 = 15 tenths. Below it, write 2.5 = ____ tenths.) Fill in the blank.
- S: 25 tenths.
- T: (Write 2.5 = 25 tenths. Below it, write 12.5 = ____ tenths.) Fill in the blank.
- S: 125 tenths.

Repeat the process for 17.5, 27.5, 24.5, 24.3, and 42.3.

Application Problems (8 minutes)

Craig, Randy, Charlie, and Sam ran in a 5K race on Saturday. They were the top 4 finishers. Here are their race times:

Craig: 25.9 minutes   Randy: 32.2 minutes   Charlie: 32.28 minutes   Sam: 25.85 minutes

Who won first place? Who won second place? Third? Fourth?

Concept Development (30 minutes)

Materials: (S) Personal white boards, place value charts, markers

Problem 1

Strategically decompose 155 using multiple units to round to the nearest ten and nearest hundred.

- T: Work with your partner and name 155 using as many hundreds as possible. Then name it using as many tens as possible, and then using as many ones as possible. Record your ideas on your place value chart.

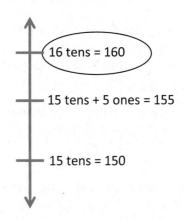

1 hundred	5 tens	5 ones
	15 tens	5 ones
		155 ones

Lesson 7: Round a given decimal to any place using place value understanding and the vertical number line.
Date: 6/28/13

1.C.3

© 2013 Common Core, Inc. All rights reserved. commoncore.org

T: Which of these decompositions of 155 helps you round this number to the nearest 10? Turn and talk.

S: 15 tens and 5 ones. The one that shows 15 tens. This helps me see that 155 is between 15 tens and 16 tens on the number line. It is exactly halfway, so 155 would round to the next greater ten which is 16 tens or 160.

T: Let's record that on the number line. (Record both nearest multiples, halfway point, number being considered, then circle rounded figure.)

T: Using your chart, which of these representations helps you round 155 to the nearest 100? Turn and talk to your partner about how you will round.

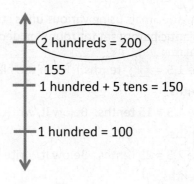

S: The one that shows 1 hundred. → I can see that 155 is between 1 hundred and 2 hundred. → The midpoint between 1 hundred and 2 hundred is 150. 155 is past the midpoint, so 155 is closer to 2 hundreds. It rounds up to 200.

T: Label your number line with the nearest multiples and then circle your rounded number.

Problem 2

Strategically decompose 1.57 to round to the nearest whole and nearest tenth.

T: Work with your partner and use your disks to name 1.57 using as many ones disks, tenths disks, and hundredths disks as possible. Write your ideas on your place value chart.

S: (Students work and share.)

1 one	5 tenths	7 hundredths
	15 tenths	7 hundredths
		157 hundredths

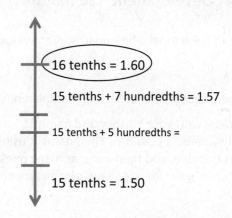

T: What decomposition of 1.57 best helps you to round this number to the nearest tenth? Turn and talk. Label your number line and circle your rounded answer.

S: (Students share.)

Bring to students' attention that this problem set parallels conversions between meters and centimeters as different units are being used to name the same quantity that is 1.57 meters = 157 centimeters.

Problem 3

Strategically decompose to round 4,381 to the nearest ten, one, tenth, and hundredth.

Lesson 7:	Round a given decimal to any place using place value understanding and the vertical number line.
Date:	6/28/13

1.C.4

COMMON CORE MATHEMATICS CURRICULUM • NY   **Lesson 7** **5•1**

T: Work with your partner and decompose 4.831 using as many tens, ones, tenths, and hundredths as possible. Record your work on your place value chart.

S: (Students share.)

0 tens	4 ones	3 tenths	8 hundredths	1 thousandth
		43 tenths	8 hundredths	1 thousandth
			438 hundredths	1 thousandth
				4381 thousandths

T: We want to round this number to the nearest 10 first. How many tens did you need to name this number?

S: No tens.

T: Between what two multiples of ten will we place this number on the number line? Turn and talk. Draw your number line and circle your rounded number.

S: (Students share.)

T: Work with your partner to round 4.381 to the nearest one, tenth, and hundredth. Explain your thinking with a number line.

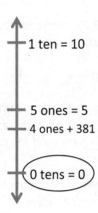

Follow the sequence from above to guide students in realizing that the number 4.381 rounds down to 4 ones, up to 44 tenths (4.4), and down to 438 hundredths (4.38).

Problem 4

Strategically decompose to round 9.975 to the nearest one, ten, tenth, and hundredth.

Follow the sequence above to lead students in rounding to the given places. This problem can prove to be a problematic rounding case. However, naming the number with different units allows students to easily choose between nearest multiples of the given place value. The decomposition chart and the number lines are given below.

0 tens	9 ones	9 tenths	7 hundredths	5 thousandths
		99 tenths	7 hundredths	5 thousandths
			997 hundredths	5 thousandths
				9975 thousandths

Lesson 7: Round a given decimal to any place using place value understanding and the vertical number line.
Date: 6/28/13

1.C.5

COMMON CORE MATHEMATICS CURRICULUM • NY   Lesson 7 5•1

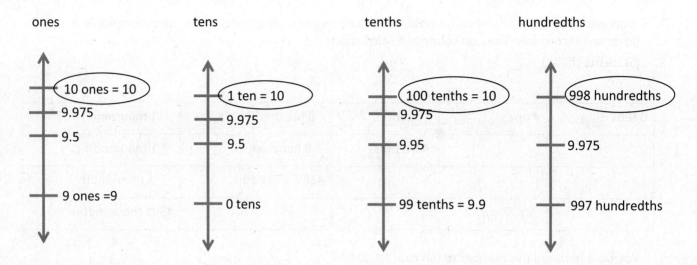

Repeat this sequence with 99.799 and round to nearest ten, one, tenth, and hundredth.

Problem Set (10 minutes)

Students should do their personal best to complete the Problem Set within the allotted 10 minutes. For some classes, it may be appropriate to modify the assignment by specifying which problems they work on first. Some problems do not specify a method for solving. Students solve these problems using the RDW approach used for Application Problems.

On this Problem Set, we suggest all students begin with Problems 1, 2, 3, and 5 and possibly leave Problem 4 to the end if they still have time.

Before circulating while students work, review the debrief questions relevant to the Problem Set so that you can better guide students to a deeper understanding of and skill with the lesson's objective.

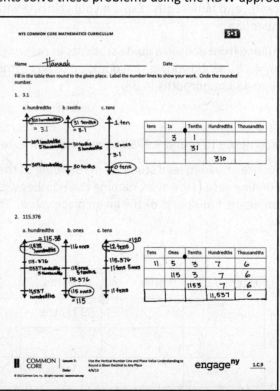

Student Debrief (10 minutes)

Lesson Objective: Round a given decimal to any place using place value understanding and the vertical number line.

The Student Debrief is intended to invite reflection and active processing of the total lesson experience.

Invite students to review their solutions for the Problem Set. They should check work by comparing answers with a partner before going over answers as a class. Look for misconceptions or misunderstandings

Lesson 7: Round a given decimal to any place using place value understanding and the vertical number line.
Date: 6/28/13

that can be addressed in the Debrief. Guide students in a conversation to debrief the Problem Set and process the lesson. You may choose to use any combination of the questions below to lead the discussion.

- In Problem 2, which decomposition helps you most if you want to round to the hundredths place? The tens place? Ones place? Why?

- How was Problem 1 different from both Problem 2 and 3? (While students may offer many differences, the salient point here is that Problem 1 is already rounded to the nearest hundredth and tenth.)

- Unit choice is the foundation of the current lesson. Problem 3 on the activity sheet offers an opportunity to discuss how the choice of unit affects the result of rounding. Be sure to allow time for these important understandings to be articulated by asking the following: If a number rounds "up" when rounded to the nearest tenth, does it follow that it will round "up" when rounded to the nearest hundredth? Thousandth? Why or why not? How do we decide about rounding "up" or "down"? How does the unit we are rounding to affect the position of the number relative to the midpoint?

- Problem 3 also offers a chance to discuss how "9" numbers often round to the same number regardless of the unit to which they are rounded. Point out that decomposing to smaller units makes this type of number easier to round because the decompositions make it simple to see which numbers are the endpoints of the segment of the number line within which the number falls.

Extension: Problem 6 offers an opportunity to discuss the effect rounding to different places has on the accuracy of a measurement. Which rounded value is closest to the actual measurement? Why? In this problem, does that difference in accuracy matter? In another situation might those differences in accuracy be more important? What should be considered when deciding to round and to which place one might round? (For some students, this may lead to an interest in significant digits and their role in measurement in other disciplines.)

Exit Ticket (3 minutes)

After the Student Debrief, instruct students to complete the Exit Ticket. A review of their work will help you assess the students' understanding of the concepts that were presented in the lesson today and plan more effectively for future lessons. You may read the questions aloud to the students.

Lesson 7: Round a given decimal to any place using place value understanding and the vertical number line.
Date: 6/28/13

A

Find the midpoint.   # Correct _____

#	a	b	#	a	b
1	0	10	23	8.5	8.6
2	0	1	24	2.8	2.9
3	0	0.01	25	0.03	0.04
4	10	20	26	0.13	0.14
5	1	2	27	0.37	0.38
6	2	3	28	80	90
7	3	4	29	90	100
8	7	8	30	8	9
9	1	2	31	9	10
10	0.1	0.2	32	0.8	0.9
11	0.2	0.3	33	0.9	1
12	0.3	0.4	34	0.08	0.09
13	0.7	0.8	35	0.09	0.1
14	0.1	0.2	36	26	27
15	0.01	0.02	37	7.8	7.9
16	0.02	0.03	38	1.26	1.27
17	0.03	0.04	39	29	30
18	0.07	0.08	40	9.9	10
19	6	7	41	7.9	8
20	16	17	42	1.59	1.6
21	38	39	43	1.79	1.8
22	0.4	0.5	44	3.99	4

© Bill Davidson

Lesson 7: Round a given decimal to any place using place value understanding and the vertical number line.
Date: 6/28/13

B Improvement _____ # Correct _____

Find the midpoint.

#	A	B	#	A	B
1	10	20	23	0.7	0.8
2	1	2	24	4.7	4.8
3	0.1	0.2	25	2.3	2.4
4	0.01	0.02	26	0.02	0.03
5	0	10	27	0.12	0.13
6	0	1	28	0.47	0.48
7	1	2	29	80	90
8	2	3	30	90	100
9	6	7	31	8	9
10	1	2	32	9	10
11	0.1	0.2	33	0.8	0.9
12	0.2	0.3	34	0.9	1
13	0.3	0.4	35	0.08	0.09
14	0.6	0.7	36	0.09	0.1
15	0.1	0.2	37	36	37
16	0.01	0.02	38	6.8	6.9
17	0.02	0.03	39	1.46	1.47
18	0.03	0.04	40	39	40
19	0.06	0.07	41	9.9	10
20	7	8	42	6.9	7
21	17	18	43	1.29	1.3
22	47	48	44	6.99	7

© Bill Davidson

Lesson 7: Round a given decimal to any place using place value understanding and the vertical number line.

Date: 6/28/13

COMMON CORE MATHEMATICS CURRICULUM • NY

Lesson 7 Problem Set 5•1

Name _____   Date _____

Fill in the table then round to the given place. Label the number lines to show your work. Circle the rounded number.

1. 3.1

   a. hundredths   b. tenths   c. tens

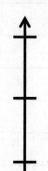
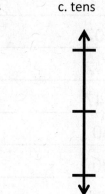

tens	1s	Tenths	Hundredths	Thousandths

2. 115.376

   a. hundredths   b. ones   c. tens

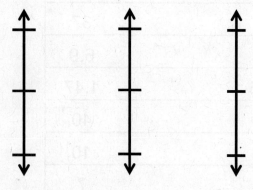

Tens	Ones	Tenths	Hundredths	Thousandths

Lesson 7: Round a given decimal to any place using place value understanding and the vertical number line.
Date: 6/28/13

3. 0.994

Tens	Ones	Tenths	Hundredths	thousandths

a. hundredths b. tenths c. ones d. tens

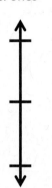

4. For open international competition, the throwing circle in the men's shot put must have a diameter of 2.135 meters. Round this number to the nearest hundredth to estimate the diameter. Use a number line to show your work.

5. Jen's pedometer said she walked 2.549 miles. She rounded her distance to 3 miles. Her brother rounded her distance to 2.5 miles. When they argued about it, their mom said they are both right. Explain how that could be true. Use number lines and words to explain your reasoning.

COMMON CORE MATHEMATICS CURRICULUM • NY   Lesson 7 Exit Ticket   5•1

Name _____   Date _____

Use the table to round the number to the given places. Label the number lines and circle the rounded value.

0	8 ones	5 tenths	4 hundredths	6 thousandths
		85 tenths	4 hundredths	6 thousandths
			854 hundredths	6 thousandths
				8546

8.546

a. hundredths   b. tens

   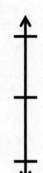

Lesson 7: Round a given decimal to any place using place value understanding and the vertical number line.
Date: 6/28/13

1.C.12

COMMON CORE MATHEMATICS CURRICULUM • NY   Lesson 7 Homework   5•1

Name _____   Date _____

Round to the given place value. Label the number lines to show your work. Circle the rounded number. Use a separate sheet to show your decompositions for each one.

1. 4.3

   a. hundredths   b. tenths   c. ones   d. tens

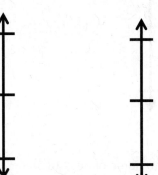
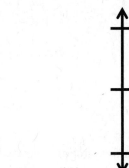
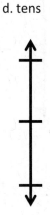

2. 225.286

   a. hundredths   b. tenths   c. ones   d. tens

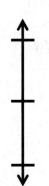

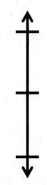

Lesson 7: Round a given decimal to any place using place value understanding and the vertical number line.
Date: 6/28/13

1.C.13

3. 8.984

   a. hundredths   b. tenths   c. ones   d. tens

4. On a major League Baseball diamond, the distance from the pitcher's mound to home plate is 18.386 meters.

   a. Round this number to the nearest hundredth of a meter to estimate the distance. Use a number line to show your work.

   b. About how many centimeters is it from the pitcher's mound to home plate?

5. Jules reads that one pint is equivalent to 0.473 liters. He asks his teacher how many liters there are in a pint. His teacher responds that there are about 0.47 liters in a pint. He asks his parents, and they say there are about 0.5 liters in a pint. Jules says they are both correct. How can that be true? Explain your answer.

Lesson 8

Objective: Round a given decimal to any place using place value understanding and the vertical number line.

Suggested Lesson Structure

■ Fluency Practice   (12 minutes)
■ Application Problems   (6 minutes)
■ Concept Development   (32 minutes)
■ Student Debrief   (10 minutes)
  Total Time   **(60 minutes)**

Fluency Practice (12 minutes)

- Rename the Units **5.NBT.3**   (6 minutes)
- Round to Different Place Values **5.NBT.4**   (6 minutes)

Rename the Units (6 minutes)

Note: Decomposing common units as decimals will strengthen student understanding of place value.

   T: (Write 13 tenths = ____.) Say the decimal.
   S: One and 3 tenths.

Repeat the process for 14 tenths, 24 tenths, 124 tenths, and 524 tenths.

   T: Name the number of tenths. (Write 2. 5 tenths.)
   S: 25 tenths.

Repeat the process for 17.5, 27.5, 24.5, 24.3, and 42.3. Repeat the entire process but with hundredths.

   T: (Write 37 hundredths = ____.) Say the decimal.
   S: 0.37
   T: (Write 37 hundredths = 0.37. Below it, write 137 hundredths = ____.) Say the decimal.
   S: 1.37

Repeat the process for 537 hundredths and 296 hundredths.

   T: (Write 0.548 = ____ thousandths.) Say the number sentence.
   S: 0.548 = 548 thousandths.

NOTES ON MULTIPLE MEANS OF ACTION AND EXPRESSION:

Learners with language differences may have more success in responding to today's sprint by writing rather than verbalizing responses. Often English language learners have receptive language abilities that exceed productive abilities, therefore allowing a choice of written response can increase their accuracy and allow for more confident participation.

Lesson 8:	Round a given decimal to any place using place value understanding and the vertical number line.
Date:	6/28/13

1.C.15

COMMON CORE MATHEMATICS CURRICULUM • NY   Lesson 8   5•1

T: (Write 0.548 = 548 thousandths. Below it, write 1.548 = ____ thousandths.) Say the number sentence.

S: 1.548 = 1548 thousandths.

Repeat the process for 2.548 and 7.352.

Round to Different Place Values (6 minutes)

Materials: (S) Personal white boards

Note: Reviewing this skill introduced in Lesson 7 will help students work towards mastery of rounding decimal numbers to different place values.

Although the approximation sign (≈) is used in Grade 4, a quick review of its meaning may be in order.

T: (Project 8.735.) Say the number.

S: 8 and 735 thousandths.

T: Draw a vertical number line on your boards with 2 endpoints and a midpoint.

T: Between what two ones is 8.735?

S: 8 ones and 9 ones.

T: What's the midpoint for 8 and 9?

S: 8.5

T: Fill in your endpoints and midpoint.

T: 8.5 is the same as how many tenths?

S: 85 tenths.

T: How many tenths are in 8.735?

S: 87 tenths.

T: (Write 8.735 ≈ _____.) Show 8.735 on your number line and write the number sentence.

S: (Students write 8.735 between 8.5 and 9 on the number line and write 8.735 ≈ 9.)

Repeat the process for the tenths place and hundredths place. Follow the same process and procedure for 7.458.

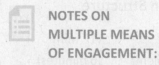

NOTES ON MULTIPLE MEANS OF ENGAGEMENT:

Turn and talk is a strategy intended to broaden active student participation by offering opportunity for all to speak during a lesson. Spend time in the beginning of the school year helping students understand what *turn and talk* looks like and sounds like by demonstrating with a student for the whole class. Modeling knee-to-knee, eye-to-eye body posture and active listening expectations (Can I restate my partner's ideas in my own words?) make for successful implementation of this powerful strategy.

Application Problem (6 minutes)

Organic, whole-wheat flour sells in bags weighing 2.915 kilograms. How much flour is this rounded to the nearest tenth? How much flour is this rounded to the nearest one? What is the difference of the two answers? Use a place value chart and number line to explain your thinking.

Lesson 8: Round a given decimal to any place using place value understanding and the vertical number line.
Date: 6/28/13

COMMON CORE MATHEMATICS CURRICULUM • NY   Lesson 8   5•1

Concept Development (32 minutes)

Materials: (S) Personal place value boards

Problem 1

Round 49.67 to the nearest ten.

T: Turn and talk to your partner about the different ways 49.67 could be decomposed using place value disks. Show the decomposition that you think will be most helpful in rounding to the nearest ten.

T: Which one of these decompositions did you decide was the most helpful?

S: The decomposition with more tens is most helpful, because it helps me identify the two rounding choices: 4 tens or 5 tens.

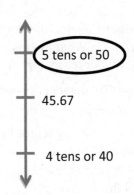

4 tens	9 ones	6 tenths	7 hundredths
	49 ones	6 tenths	7 hundredths
		496 tenths	7 hundredths

T: Draw and label a number line and circle the rounded value. Explain your reasoning.

Repeat this sequence with rounding 49.67 to the nearest ones, and then tenths.

Problem 2

Decompose 9.949 and round to the nearest tenth and hundredth. Show your work on a number line.

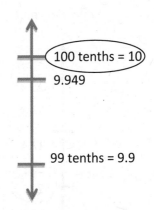

9 ones	9 tenths	4 hundredths	9 thousandths
	99 tenths	4 hundredths	9 thousandths
		994 hundredths	9 thousandths

T: What decomposition of 9.949 best helps to round this number to the nearest tenth?

S: The one using the most tenths to name the decimal fraction. I knew I would round to either 99 tenths or 100 tenths. I looked at the hundredths. Nine hundredths is past the midpoint, so I rounded to the next tenth, 100 tenths. One hundred tenths is the same as 10.

Lesson 8: Round a given decimal to any place using place value understanding and the vertical number line.
Date: 6/28/13

1.C.17

T: Which digit made no difference when you rounded to the nearest tenth? Explain your thinking.

S: The thousandths, because the hundredths decided which direction to round. As long as I had 5 hundredths, I was past the halfway point so I rounded to the next number.

Repeat the process rounding to the nearest hundredth.

Problem 3

A decimal number has 1 digit to the right of the decimal point. If we round this number to the nearest whole number, the result is 27. What are the maximum and minimum possible values of these two numbers? Use a number line to show your reasoning. Include the midpoint on the number line.

T: (Draw a vertical number line with 3 points.)

T: What do we know about the unknown number?

S: It has a number in the tenths place, but nothing else past the decimal point. We know that is has been rounded to 27.

T: (Write 27 at the bottom point on the number line and circle it.) Why did I place 27 as the lesser rounded value?

S: We are looking for the largest number that will round down to 27. That number will be greater than 27, but less than the midpoint between 27 and 28.

T: What is the midpoint between 27 and 28?

S: 27.5

T: (Place 27.5 on the number line.)

T: If we look at numbers that have exactly 1 digit to the right of the decimal point, what is the greatest one that will round down to 27?

S: 27.4. If we go to 27.5, that would round up to 28.

Repeat the same process to find the minimum value.

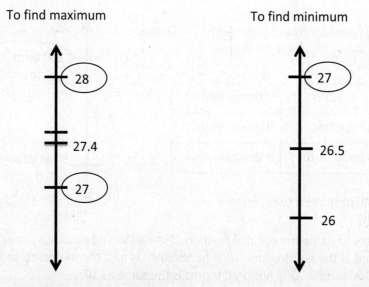

Encourage further discussion with the following:

What if our number had exactly 2 digits to the right of the decimal point? Could I find a number larger than 27.4 that would still round down to 27? (Various answers could be expected: 27.41, 27.49, etc.). What is the largest possible value it could have? (27.49.)

A similar discussion can take place in finding the minimum when students discover that 26.5 rounds up to 27. Lead students to discover that something different happens here. Can we find a number less than 26.5 with exactly 2 digits to the right of the decimal point that would still round up? (No, nothing smaller than 26.50.)

Problem Set (10 minutes)

Students should do their personal best to complete the problem set within the allotted 10 minutes. For some classes, it may be appropriate to modify the assignment by specifying which problems they work on first. Some problems do not specify a method for solving. Students solve these problems using the RDW approach used for Application Problems.

On this Problem Set, we suggest all students begin with Problems 1 and 3 and possibly leave Problem 2 to the end if they still have time.

Before circulating while students work, review the debrief questions relevant to the problem set so that you can better guide students to a deeper understanding of a skill with the lesson's objective.

Student Debrief (10 minutes)

Lesson Objective: Round a given decimal to any place using place value understanding and the vertical number line.

The Student Debrief is intended to invite reflection and active processing of the total lesson experience.

Invite students to review their solutions for the Problem Set. They should check work by comparing answers with a partner before going over answers as a class. Look for misconceptions or misunderstandings that can be addressed in the Debrief. Guide students in a conversation to debrief the Problem Set and process the lesson. You may choose to use any combination of the questions below to lead the discussion.

- Compare our approach to rounding today and yesterday. How are they alike? How are they different? (Students will likely offer many accurate responses. However, lead the discussion toward the notion of our only choosing specific decompositions to round in today's lesson as opposed to naming every

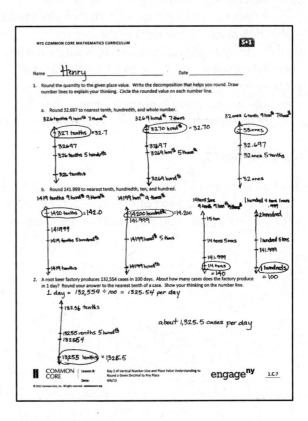

decomposition in yesterday's lesson. Also explore which units (place values) are worthy of attention and which are not when rounding to a specific place value. Are there patterns to these choices?)

- Once a number rounds up at one place value, does it follow then that every place value will round up? Why or why not? (Encourage students to reference their problem sets as evidence of their reasoning. Problem 1(b) provides an example of differing unit choices resulting in differences in rounding up and down.)
- How does the place value chart help organize your thinking when rounding?
- Finding the maximum and minimum values poses a significant increase in cognitive load and an opportunity to build excitement! Make time to deeply discuss ways of reasoning about these tasks, as they are sure to be many and varied. Consider a discussion of Problem 3 that mirrors the one in the lesson: What if our number had exactly three digits to the right of the decimal?

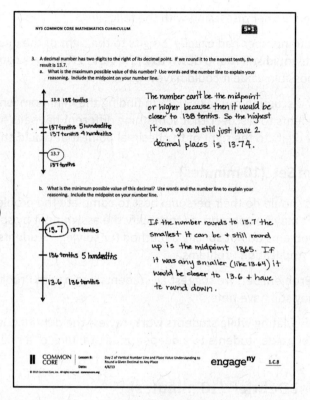

Can we find a value larger than 13.74 that would round down to 13.7? (13.749) What about 4 places or 5 places to the right of the decimal? (13.7499, 13.74999) Encourage students to generalize that we can get infinitely close to 13.5 with a decimal that has an infinite number of 9's yet that decimal will still round down to 13.7. We can find points on the number line as close as we like, and yet they will not be equal to 13.75. Follow that with the discovery that this is not true for our minimum value. There is nothing smaller than 13.750 that will round up to 13.8. Math journals offer a venue for students to continue to explore maximum and minimum tasks beyond today's lesson.

Exit Ticket (3 minutes)

After the Student Debrief, instruct students to complete the Exit Ticket. A review of their work will help you assess the students' understanding of the concepts that were presented in the lesson today and plan more effectively for future lessons. You may read the questions aloud to the students.

COMMON CORE MATHEMATICS CURRICULUM • NY Lesson 8 Problem Set 5•1

Name _____ Date _____

1. Write the decomposition that helps you, and then round to the given place value. Draw number lines to explain your thinking. Circle the rounded value on each number line.

   a. Round 32.697 to nearest tenth, hundredth, and whole number.

   b. Round 141.999 to nearest tenth, hundredth, ten, and hundred.

2. A root beer factory produces 132,554 cases in 100 days. About how many cases does the factory produce in 1 day? Round your answer to the nearest tenth of a case. Show your thinking on the number line.

Lesson 8: Round a given decimal to any place using place value understanding and the vertical number line.
Date: 6/28/13

3. A decimal number has two digits to the right of its decimal point. If we round it to the nearest tenth, the result is 13.7.

   a. What is the maximum possible value of this number? Use words and the number line to explain your reasoning. Include the midpoint on your number line.

   b. What is the minimum possible value of this decimal? Use words and the number line to explain your reasoning. Include the midpoint on your number line.

COMMON CORE MATHEMATICS CURRICULUM • NY   Lesson 8 Exit Ticket   5•1

Name _____   Date _____

1. Round the quantity to the given place value. Draw number lines to explain your thinking. Circle the rounded value on the number line.

   a. 13.989 to nearest tenth

   b. 382.993 to nearest hundredth

COMMON CORE MATHEMATICS CURRICULUM • NY   Lesson 8 Homework   5•1

Name _____   Date _____

1. Round the quantity to the given place value. Draw number lines to explain your thinking. Circle the rounded value on the number line.

   a. 43.586 to nearest tenth, hundredth, and whole number

   b. 243.875 to nearest tenth, hundredth, ten, and hundred

2. A trip from New York City to Seattle is 2,852.1 miles. A family wants to make the drive in 10 days, driving the same number of miles each day. About how many miles will they drive each day? Round you answer to the nearest tenth of a mile.

Lesson 8: Round a given decimal to any place using place value understanding and the vertical number line.
Date: 6/28/13

1.C.24

3. A decimal number has two digits to the right of its decimal point. If we round it to the nearest tenth, the result is 18.6.

   a. What is the maximum possible value of this decimal? Use words and the number line to explain your reasoning.

   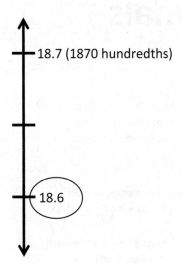

   b. What is the minimum possible value of this decimal? Use words, numbers and pictures to explain your reasoning.

   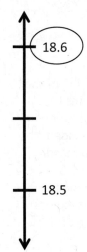

COMMON CORE MATHEMATICS CURRICULUM • NY   Topic D **5•1**

GRADE 5 • MODULE 1

Topic D
Adding and Subtracting Decimals

5.NBT.2, 5.NBT.3, 5.NBT.7

Focus Standard:	5.NBT.2	Explain patterns in the number of zeros of the product when multiplying a number by powers of 10, and explain patterns in the placement of the decimal point when a decimal is multiplied or divided by a power of 10. Use whole-number exponents to denote powers of 10.
	5.NBT.3	Read, write, and compare decimals to thousandths.
		a. Read and write decimals to thousandths using base-ten numerals, number names, and expanded form, e.g., 347.392 = 3 × 100 + 4 × 10 + 7 × 1 + 3 × (1/10) + 9 × (1/100) + 2 × (1/1000).
		b. Compare two decimals to thousandths based on meanings of the digits in each place, using >, =, and < symbols to record the results of comparisons.
	5.NBT.7	Add, subtract, multiply and divide decimals to hundredths, using concrete models or drawings and strategies based on place value, properties of operations, and/or the relationship between addition and subtraction; relate the strategy to a written method and explain the reasoning used.
Instructional Days:	2	
Coherence -Links from:	G4–M1	Place Value, Rounding, and Algorithms for Addition and Subtraction
-Links to:	G6–M2	Arithmetic Operations Including Dividing by a Fraction

Topics D through F mark a shift from the opening topics of Module 1. From this point to the conclusion of the module, students begin to use base ten understanding of adjacent units and whole number algorithms to reason about and perform decimal fraction operations—addition and subtraction in Topic D, multiplication in Topic E and division in Topic F (**5.NBT.7**). In Topic D, unit form provides the connection that allows students to use what they know about general methods for addition and subtraction with whole numbers to reason about decimal addition and subtraction, e.g., 7 tens + 8 tens = 15 tens = 150 is analogous to 7 tenths + 8 tenths = 15 tenths = 1.5. Place value charts and disks (both concrete and pictorial representations) and the relationship between addition and subtraction are used to provide a bridge for relating such understandings to a written method. Real world contexts provide opportunity for students to apply their knowledge of decimal addition and subtraction as well in Topic D.

Topic D

A Teaching Sequence Towards Mastery of Adding and Subtracting Decimals

Objective 1: Add decimals using place value strategies and relate those strategies to a written method.
(Lesson 9)

Objective 2: Subtract decimals using place value strategies and relate those strategies to a written method.
(Lesson 10)

Lesson 9

Objective: Add decimals using place value strategies and relate those strategies to a written method.

Suggested Lesson Structure

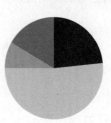

- ■ Fluency Practice   (14 minutes)
- ■ Application Problems   (5 minutes)
- ■ Concept Development   (31 minutes)
- ■ Student Debrief   (10 minutes)
- **Total Time**   **(60 minutes)**

Fluency Practice (14 minutes)

- Round to the Nearest One **5.NBT.4**   (8 minutes)
- Decompose the Unit **5.NBT.1**   (2 minutes)
- Round to Different Place Values **5.NBT.4**   (2 minutes)
- One Unit More **5.NBT.7**   (2 minutes)

Sprint: Round to the Nearest One (8 minutes)

Materials: (S) Round to the Nearest One Sprint

Note: This Sprint will help students build mastery of rounding to the nearest whole number.

Decompose the Unit (2 minutes)

Materials: (S) Personal white boards

Note: Decomposing common units as decimals will strengthen student understanding of place value.

- T: (Project 6.358.) Say the number.
- S: 6 and 358 thousandths.
- T: How many tenths are in 6.358?
- S: 63 tenths.
- T: (Write 6.358 = 63 tenths ____ hundredths.) On your boards, write the number separating the tenths.
- S: (Students write 6.358 = 63 tenths 58 thousandths.)

Repeat process for hundredths. Follow the same process for 7.354.

Lesson 9: Add decimals using place value strategies and relate those strategies to a written method.
Date: 6/28/13

COMMON CORE MATHEMATICS CURRICULUM • NY   **Lesson 9** **5•1**

Round to Different Place Values (2 minutes)

Materials: (S) Personal white boards

Note: Reviewing this skill that was introduced in lesson 8 will help students work towards mastery of rounding decimal numbers to different place values.

- T: (Project 2.475.) Say the number.
- S: 2 and 475 thousandths.
- T: On your boards, round the number to the nearest tenth.

Students write 2.475 ≈ 2.5. Repeat the process, rounding 2.457 to the nearest hundredth. Follow the same process, but vary the sequence for 2.987.

One Unit More (2 minutes)

Materials: (S) Personal white boards

Note: This anticipatory fluency drill will lay a foundation for the concept taught in this lesson.

- T: (Write 5 tenths.) Say the decimal that's one tenth more than the given value.
- S: 0.6

Repeat the process for 5 hundredths, 5 thousandths, 8 hundredths, 3 tenths, and 2 thousandths. Specify the unit to increase by.

- T: (Write 0.052.) On your board, write one more thousandth.
- S: 0.053

Repeat the process for 1 tenth more than 35 hundredths, 1 thousandth more than 35 hundredths, and 1 hundredth more than 438 thousandths.

Application Problems (5 minutes)

Ten baseballs weigh 1,417.4 grams. About how much does 1 baseball weigh? Round your answer to the nearest tenth of a gram. Round your answer to the nearest gram. If someone asked you, "About how much does a baseball weigh?" which answer would you give? Why?

Note: The application problem requires students to use skills learned in the first part of this module: dividing by powers of ten, and rounding.

Lesson 9:	Add decimals using place value strategies and relate those strategies to a written method.
Date:	6/28/13

1.D.4

COMMON CORE MATHEMATICS CURRICULUM • NY   Lesson 9 5•1

Concept Development (31 minutes)

Materials: (S) Place value chart, place value disks

Problems 1–3

2 tenths + 6 tenths

2 ones 3 thousandths + 6 ones 1 thousandth

2 tenths 5 thousandths + 6 hundredths

T: Solve 2 tenths plus 6 tenths using disks on your place value chart. (Write 2 tenths + 6 tenths on the board.)

S: (Students solve.)

T: Say the sentence in words.

S: 2 tenths + 6 tenths = 8 tenths.

T: How is this addition problem the same as a whole number addition problem? Turn and share with your partner.

S: In order to find the sum, I added like units – tenths with tenths. → 2 tenths plus 6 tenths equals 8 tenths just like 2 apples plus 6 apples equals 8 apples. → Since the sum is 8 tenths, we don't need to bundle or regroup.

T: Work with your partner and solve the next two problems with disks on your place value chart.

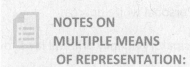

NOTES ON MULTIPLE MEANS OF REPRESENTATION:

Understanding the meaning of *tenths*, *hundredths*, and *thousandths* is essential. Proportional manipulatives, such as base ten blocks, can be used to ensure understanding of the vocabulary. Students should eventually move to concrete number disks and/or drawing, which are more efficient.

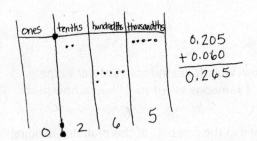

S: (Students solve.)

T: Let's record our last problem vertically. (Write 0.205 and the plus sign underneath on board.) What do I need to think about when I write my second addend?

Lead students to see that the vertical written method mirrors the placement of disks on the chart. Like units should be aligned with like units. Avoid procedural language like *line up the decimals*. Students should justify alignment of digits based on place value units.

Lesson 9: Add decimals using place value strategies and relate those strategies to a written method.
Date: 6/28/13

1.D.5

Problems 4–6

1.8 + 13 tenths

1 hundred 8 hundredths + 2 ones 4 hundredths

148 thousandths + 7 ones 13 thousandths

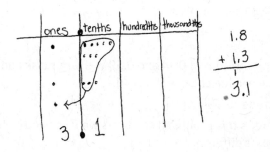

- T: Use your place value chart and disks to show the addends of our next problem. (Write "1.8 + 13 tenths" horizontally on the board.)
- S: (Students show.)
- T: Tell how you represented these addends. (Students may represent 13 tenths using 13 tenth disks or as 1 one disk and 3 tenths disks. Others may represent 1.8 using mixed units or only tenths.)
- S: (Students share.)
- T: Which way of composing these addends requires the least amount of drawing? Why?
- S: Using ones and tenths because drawing 1 one disk is faster than drawing 10 tenths.
- T: Will your choice of units in your drawing affect your answer (sum)?
- S: No! Either drawing is OK. It will still give the same answer.
- T: Add. Share your thinking with your partner.
- S: 1.8 + 13 tenths = 1 and 21 tenths. There are 10 tenths in one whole. I can compose 2 wholes and 11 tenths from 21 tenths, so the answer is 3 and 1 tenth. → 13 tenths is the same as 1 one 3 tenths. 1 one 3 tenths + 1 one 8 tenths = 2 ones 11 tenths which is the same as 3 ones 1 tenth.
- T: Let's record what we did on our charts. (Lead students to articulate the alignment of digits in the vertical equation based on like units.)
- T: What do you notice that was different about this problem? What was the same? Turn and talk.
- S: We needed to rename in this problem because 8 tenths and 3 tenths is 11 tenths. → We added ones with ones and tenths with tenths – like units just like before.
- T: Work with your partner and solve the next two problems on your place value chart and record your thinking vertically.

(As students work 148 thousandths + 7 ones 13 thousandths, discuss which composition of 148 thousandths is the more efficient for drawing on a mat.)

MULTIPLE MEANS OF ACTION AND EXPRESSION:

Some students may struggle when asked to turn and talk to another student because they need more time to compose their thoughts. Math journals can be used in conjunction with Turn and Talk as journals provide a venue in which students can use a combination of graphics, symbols and words to help them communicate their thinking.

Lesson 9: Add decimals using place value strategies and relate those strategies to a written method.
Date: 6/28/13

COMMON CORE MATHEMATICS CURRICULUM • NY

Lesson 9   5•1

Problems 7–9

0.74 + 0.59

7.048 + 5.196

7.44 + 0.774

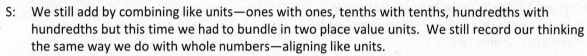

T: Find the sum of 0.74 and 0.59 with your disks on your place value chart and record.

S: (Students solve.)

T: How is this problem like others we've solved? How was it different?

S: We still add by combining like units—ones with ones, tenths with tenths, hundredths with hundredths but this time we had to bundle in two place value units. We still record our thinking the same way we do with whole numbers—aligning like units.

T: Solve the next two problems using the written method. You may also use your disks to help you. (Show 7.048 + 5.196 and 7.44 + 0.704 on the board.)

S: (Students solve.)

T: How is 7.44 + 0.704 different from the other problems we've worked? Turn and talk.

S: One addend had hundredths, the other had thousandths, but we still had to add like units. → We could think of 44 hundredths as 440 thousandths. → One addend did not have a zero in the ones place. I could leave it like that, or include the zero. The missing zero did not change the quantity.

Problem Set (10 minutes)

Students should do their personal best to complete the Problem Set within the allotted 10 minutes. For some classes, it may be appropriate to modify the assignment by specifying which problems they work on first. Some problems do not specify a method for solving. Students solve these problems using the RDW approach used for Application Problems..

On this Problem Set, we suggest all students work directly through all problems. Please note that Problem 4 includes the word *pedometer* which may need explanation for some students.

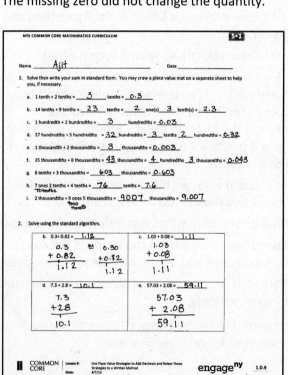

Student Debrief (10 minutes)

Lesson Objective: Add decimals using place value strategies and relate those strategies to a written method.

The Student Debrief is intended to invite reflection and active processing of the total lesson experience.

Lesson 9:   Add decimals using place value strategies and relate those strategies
            to a written method.
Date: 6/28/13

1.D.7

Invite students to review their solutions for the Problem Set. They should check work by comparing answers with a partner before going over answers as a class. Look for misconceptions or misunderstandings that can be addressed in the Debrief. Guide students in a conversation to debrief the Problem Set and process the lesson. You may choose to use any combination of the questions below to lead the discussion.

- How is adding decimal fractions the same as adding whole numbers? How is it different?
- What are some different words you have used through the grades for changing 10 smaller units for 1 of the next larger units or changing 1 unit for 10 of the next smaller units?
- What do you notice about the addends in letters (b), (d), and (f) in Problem 1? Explain the thought process in solving these problems.
- Did you recognize a pattern in the digits used in Problem 2? Look at each row and column.
- What do you notice about the sum in Problem 2(f)? What are some different ways to express the sum? (Encourage students to name the sum using thousandths, hundredths, and tenths.) How is this problem different from adding whole numbers?
- Ask early finishers to generate addition problems which have 2 decimal place values, but add up to specific sums, like 1 or 2 (e.g., 0.74 + 0.26).

Exit Ticket (3 minutes)

After the Student Debrief, instruct students to complete the Exit Ticket. A review of their work will help you assess the students' understanding of the concepts that were presented in the lesson today and plan more effectively for future lessons. You may read the questions aloud to the students.

Lesson 9: Add decimals using place value strategies and relate those strategies to a written method.
Date: 6/28/13

A # Correct _____

Round to the nearest whole number.

#	Problem		#	Problem	
1	3.1 ≈		23	12.51 ≈	
2	3.2 ≈		24	16.61 ≈	
3	3.3 ≈		25	17.41 ≈	
4	3.4 ≈		26	11.51 ≈	
5	3.5 ≈		27	11.49 ≈	
6	3.6 ≈		28	13.49 ≈	
7	3.9 ≈		29	13.51 ≈	
8	13.9 ≈		30	15.51 ≈	
9	13.1 ≈		31	15.49 ≈	
10	13.5 ≈		32	6.3 ≈	
11	7.5 ≈		33	7.6 ≈	
12	8.5 ≈		34	49.5 ≈	
13	9.5 ≈		35	3.45 ≈	
14	19.5 ≈		36	17.46 ≈	
15	29.5 ≈		37	11.76 ≈	
16	89.5 ≈		38	5.2 ≈	
17	2.4 ≈		39	12.8 ≈	
18	2.41 ≈		40	59.5 ≈	
19	2.42 ≈		41	5.45 ≈	
20	2.45 ≈		42	19.47 ≈	
21	2.49 ≈		43	19.87 ≈	
22	2.51 ≈		44	69.51 ≈	

© Bill Davidson

Lesson 9: Add decimals using place value strategies and relate those strategies to a written method.
Date: 6/28/13

B Improvement _____   # Correct _____

Round to the nearest whole number.

#	Problem	Answer	#	Problem	Answer
1	4.1 ≈		23	13.51 ≈	
2	4.2 ≈		24	17.61 ≈	
3	4.3 ≈		25	18.41 ≈	
4	4.4 ≈		26	12.51 ≈	
5	4.5 ≈		27	12.49 ≈	
6	4.6 ≈		28	14.49 ≈	
7	4.9 ≈		29	14.51 ≈	
8	14.9 ≈		30	16.51 ≈	
9	14.1 ≈		31	16.49 ≈	
10	14.5 ≈		32	7.3 ≈	
11	7.5 ≈		33	8.6 ≈	
12	8.5 ≈		34	39.5 ≈	
13	9.5 ≈		35	4.45 ≈	
14	19.5 ≈		36	18.46 ≈	
15	29.5 ≈		37	12.76 ≈	
16	79.5 ≈		38	6.2 ≈	
17	3.4 ≈		39	13.8 ≈	
18	3.41 ≈		40	49.5 ≈	
19	3.42 ≈		41	6.45 ≈	
20	3.45 ≈		42	19.48 ≈	
21	3.49 ≈		43	19.78 ≈	
22	3.51 ≈		44	59.51 ≈	

© Bill Davidson

COMMON CORE MATHEMATICS CURRICULUM • NY   Lesson 9 Problem Set   5•1

Name _____   Date _____

1. Solve then write your sum in standard form. You may draw a place value mat on a separate sheet to help you, if necessary.

   a. 1 tenth + 2 tenths = _____ tenths = _____

   b. 14 tenths + 9 tenths = _____ tenths = _____ one(s) _____ tenth(s) = _____

   c. 1 hundredth + 2 hundredths = _____ hundredths = _____

   d. 27 hundredths + 5 hundredths = _____ hundredths = _____ tenths _____ hundredths = _____

   e. 1 thousandth + 2 thousandths = _____ thousandths = _____

   f. 35 thousandths + 8 thousandths = ____ thousandths = ____ hundredths ____ thousandths = _____

   g. 6 tenths + 3 thousandths = _____ thousandths = _____

   h. 7 ones 2 tenths + 4 tenths = _____ tenths = _____

   i. 2 thousandths + 9 ones 5 thousandths = _____ thousandths = _____

2. Solve using the standard algorithm.

a. 0.3 + 0.82 = _____	b. 1.03 + 0.08 = _____
c. 7.3 + 2.8 = _____	d. 57.03 + 2.08 = _____

Lesson 9: Add decimals using place value strategies and relate those strategies to a written method.
Date: 6/28/13

e. 62.573 + 4.328 = _____	f. 85.703 + 12.197 = _____

3. Van Cortlandt Park's walking trail is 1.02 km longer than Marine Park. Central Park's walking trail is 0.242 km longer than Van Cortlandt's.

   a. Fill in the missing information in the chart below.

New York City Walking Trails	
Central Park	_____ km
Marine Park	1.28 km
Van Cortlandt Park	_____ km

   b. If a tourist walked all 3 trails in a day, how many km would they have walked?

4. Meyer has 0.64 GB of space remaining on his iPod. He wants to download a pedometer app (0.24 GB) a photo app (0.403 GB) and a math app (0.3 GB). Which combinations of apps can he download? Explain your thinking.

Lesson 9: Add decimals using place value strategies and relate those strategies to a written method.
Date: 6/28/13

COMMON CORE MATHEMATICS CURRICULUM • NY   Lesson 9 Exit Ticket   5•1

Name _____   Date _____

1. Solve.
    a. 4 hundredths + 8 hundredths = _____ hundredths = _____ tenths _____ hundredths

    b. 64 hundredths + 8 hundredths = _____ hundredths = _____ tenths _____ hundredths

2. Solve using the standard algorithm.

a. 2.40 + 1.8 = _____	b. 36.25 + 8.67 = _____

Lesson 9: Add decimals using place value strategies and relate those strategies to a written method.
Date: 6/28/13

Name _____   Date _____

1. Solve.

   a. 3 tenths + 4 tenths = _____ tenths

   b. 12 tenths + 9 tenths = _____ tenths = _____ one(s) _____ tenth(s)

   c. 3 hundredths + 4 hundredths = _____ hundredths

   d. 27 hundredths + 7 hundredths = _____ hundredths = _____ tenths _____ hundredths

   e. 4 thousandth + 3 thousandths = _____ thousandths

   f. 39 thousandths + 5 thousandths = _____ thousandths = _____ hundredths _____ thousandths

   g. 5 tenths + 7 thousandths = _____ thousandths

   h. 4 ones 4 tenths + 4 tenths = _____ tenths

   i. 8 thousandths + 6 ones 8 thousandths = _____ thousandths

2. Solve using the standard algorithm.

a. 0.4 + 0.7 = _____	b. 2.04 + 0.07 = _____
c. 6.4 + 3.7 = _____	d. 56.04 + 3.07 = _____

Lesson 9: Add decimals using place value strategies and relate those strategies to a written method.

| e. 72.564 + 5.137 = _____ | f. 75.604 + 22.296 = _____ |

3. Walkway Over the Hudson, a bridge that crosses the Hudson River in Poughkeepsie, is 2.063 kilometers. Anping Bridge, which was built in China 850 years ago, is 2.07 kilometers long.

   a. Which bridge is longer? How much longer? Show your thinking.

   b. Leah likes to walk her dog on the Walkway Over the Hudson. If she walks across and back, how far do she and her dog walk?

4. For his parents' anniversary, Danny spends $5.87 on a photo. He also buys 3 balloons for $2.49 each and a box of strawberries for $4.50. How much money does he spend all together?

Lesson 9: Add decimals using place value strategies and relate those strategies to a written method.
Date: 6/28/13

COMMON CORE MATHEMATICS CURRICULUM • NY   Lesson 10   5•1

Lesson 10

Objective: Subtract decimals using place value strategies and relate those strategies to a written method.

Suggested Lesson Structure

- ■ Fluency Practice   (10 minutes)
- ■ Application Problems   (5 minutes)
- ■ Concept Development   (35 minutes)
- ■ Student Debrief   (10 minutes)
- **Total Time**   **(60 minutes)**

Fluency Practice (10 minutes)

- Take Out the Unit **5.NBT.1**   (3 minutes)
- Add Decimals **5.NBT.7**   (3 minutes)
- One Less Unit **5.NBT.7**   (4 minutes)

Take Out the Unit (3 minutes)

Materials: (S) Personal white boards

Note: Decomposing common units as decimals will strengthen student understanding of place value.

- T: (Project 76.358 = _____.) Say the number.
- S: 76 and 358 thousandths.
- T: (Write 76.358 = 7 tens _____ thousandths.) On your board, fill in the blank.
- S: (Students write 76.358 = 7 tens 6358 thousandths.)

Repeat the process for tenths and hundredths 76.358 = 763 tenths _____ thousandths, 76.358 = ____ hundredths 8 thousandths.

Add Decimals (3 minutes)

Materials: (S) Personal white boards

Note: Reviewing this skill that was introduced in Lesson 9 will help students work towards mastery of adding common decimal units.

- T: (Write 3 tenths + 2 tenths = _____.) Write the addition sentence in decimal form.
- S: 0.3 + 0.2 = 0.5

| Lesson 10: | Subtract decimals using place value strategies and relate those strategies to a written method. |
| Date: | 6/28/13 |

1.D.16

© 2013 Common Core, Inc. All rights reserved. commoncore.org

Repeat the process for 5 hundredths + 4 hundredths and 35 hundredths + 4 hundredths.

One Unit Less (4 minutes)

Materials: (S) Personal white boards

Note: This anticipatory fluency drill will lay a foundation for the concept taught in this lesson.

T: (Write 5 tenths.) Say the decimal that is 1 less than the given unit.
S: 0.4

Repeat the process for 5 hundredths, 5 thousandths, 7 hundredths, and 9 tenths.

T: (Write 0.029.) On your board, write the decimal that is one less thousandth.
S: 0.028

Repeat the process for 1 tenth less than 0.61, 1 thousandth less than 0.061, and 1 hundredth less than 0.549.

Note: *Add Decimals* is a review of skills learned in Lesson 9. The discussion of adding like units provides a bridge to the subtraction of like units which is the topic of today's lesson.

Application Problems (5 minutes)

At the 2012 London Olympics, Michael Phelps won the gold medal in the men's 100 meter butterfly. He swam the first 50 meters in 26.96 seconds. The second 50 meters took him 25.39 seconds. What was his total time?

Concept Development (35 minutes)

Materials: (S) Place value chart, personal white boards, markers per student

Problem 1

5 tenths – 3 tenths

7 ones 5 thousandths – 2 ones 3 thousandths

9 hundreds 5 hundredths – 3 hundredths

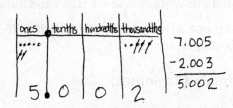

T: (Write 5 tenths – 3 tenths = _____ on the board.) Let's read this expression aloud together. Turn and tell your partner how you'll solve this problem, then find the difference using your place value chart.

T: Explain your reasoning when solving this subtraction sentence.

S: Since the units are alike we can just subtract. 5 – 3 = 2.
→ This problem is very similar to 5 ones minus 2 ones, 1 or 5 people minus 2 people; the units may change

3. Solve.

a. 10 tens − 1 ten 1 tenth	b. 3 − 22 tenths	c. 37 tenths − 1 one 2 tenths
d. 8 ones 9 hundredths − 3.4	e. 5.622 − 3 hundredths	f. 2 ones 4 tenths − 0.59

4. Mrs. Fan wrote 5 tenths minus 3 hundredths on the board. Michael said the answer is 2 tenths because 5 minus 3 is 2. Is he correct? Explain.

5. A pen costs $2.09. It costs $0.45 less than a marker. Ken paid for one pen and one marker with a five dollar bill. Use a tape diagram with calculations to determine his change.

COMMON CORE MATHEMATICS CURRICULUM • NY   Lesson 10 Exit Ticket   5•1

Name _____   Date _____

1. Subtract.

    1.7 – 0.8 = _____ tenths – _____ tenths = _____ tenths = _____

2. Subtract vertically, showing all work.

    a. 84.637 – 28.56 = _____

    b. 7 – 0.35 = _____

Lesson 10: Subtract decimals using place value strategies and relate those strategies to a written method.
Date: 6/28/13

COMMON CORE MATHEMATICS CURRICULUM • NY   Lesson 10 Homework   5•1

Name _____   Date _____

1. Subtract. You may use a place value chart.

   a. 9 tenths – 3 tenths = _____ tenth

   b. 9 ones 2 thousandths – 3 ones = _____ ones _____ thousandths

   c. 4 hundreds 6 hundredths – 3 hundredths = _____ hundreds _____ hundredths

   d. 56 thousandths – 23 thousandths = _____ thousandths

                                      = _____ hundredths _____ thousandths

2. Solve using the standard algorithm.

a. 1.8 – 0.9 = _____	b. 41.84 – 0.9 = _____	c. 341.84 – 21.92 = _____
d. 5.182 – 0.09 = _____	e. 50.416 – 4.25 = _____	f. 741. – 3.91 = _____

Lesson 10: Subtract decimals using place value strategies and relate those strategies to a written method.

3. Solve.

a. 30 tens – 3 tens 3 tenths	b. 5 – 16 tenths	c. 24 tenths – 1 one 3 tenths
d. 6 ones 7 hundredths – 2.3	e. 8.246 – 5 hundredths	f. 5 ones 3 tenths – 0.53

4. Mr. House wrote 8 tenths minus 5 hundredths on the board. Maggie said the answer is 3 hundredths because 8 minus 5 is 3. Is she correct? Explain.

5. A clipboard costs $2.23. It costs $0.58 more than a notebook. Lisa buys two clipboards and one notebook, and paid with a ten dollar bill. Use a tape diagram with calculations to show her change.

GRADE 5 • MODULE 1

Topic E
Multiplying Decimals

5.NBT.2, 5.NBT.3, 5.NBT.7

Focus Standard:	5.NBT.2	Explain patterns in the number of zeros of the product when multiplying a number by powers of 10, and explain patterns in the placement of the decimal point when a decimal is multiplied or divided by a power of 10. Use whole-number exponents to denote powers of 10.
	5.NBT.3	Read, write, and compare decimals to thousandths.
		a. Read and write decimals to thousandths using base-ten numerals, number names, and expanded form, e.g., 347.392 = 3 × 100 + 4 × 10 + 7 × 1 + 3 × (1/10) + 9 × (1/100) + 2 × (1/1000).
		b. Compare two decimals to thousandths based on meanings of the digits in each place, using >, =, and < symbols to record the results of comparisons.
	5.NBT.7	Add, subtract, multiply and divide decimals to hundredths, using concrete models or drawings and strategies based on place value, properties of operations, and/or the relationship between addition and subtraction; relate the strategy to a written method and explain the reasoning used.
Instructional Days:	2	
Coherence -Links from:	G4–M3	Multi-Digit Multiplication and Division
-Links to:	G5–M2	Multi-Digit Whole Number and Decimal Fraction Operations
	G6–M2	Arithmetic Operations Including Dividing by a Fraction

A focus on reasoning about the multiplication of a decimal fraction by a one-digit whole number in Topic E provides the link that connects Grade 4 multiplication work and Grade 5 fluency with multi-digit multiplication. Place value understanding of whole number multiplication coupled with an area model of the distributive property is used to help students build direct parallels between whole number products and the products of one-digit multipliers and decimals (**5.NBT.7**). Students use an estimation based strategy to confirm the reasonableness of the product once the decimal has been placed through place value reasoning. Word problems provide a context within which students can reason about products.

COMMON CORE MATHEMATICS CURRICULUM • NY   Topic E 5•1

A Teaching Sequence Towards Mastery of Multiplying Decimals

Objective 1: Multiply a decimal fraction by single-digit whole numbers, relate to a written method through application of the area model and place value understanding, and explain the reasoning used.
(Lesson 11)

Objective 2: Multiply a decimal fraction by single-digit whole numbers, including using estimation to confirm the placement of the decimal point.
(Lesson 12)

| Topic E: | Multiplying Decimals |
| Date: | 6/28/13 |

© 2013 Common Core, Inc. All rights reserved. commoncore.org

COMMON CORE MATHEMATICS CURRICULUM • NY   Lesson 11   5•1

Lesson 11

Objective: Multiply a decimal fraction by single-digit whole numbers, relate to a written method through application of the area model and place value understanding, and explain the reasoning used.

Suggested Lesson Structure

- **Fluency Practice**   (10 minutes)
- **Application Problems** (5 minutes)
- **Concept Development**   (35 minutes)
- **Student Debrief**   (10 minutes)
- **Total Time**   **(60 minutes)**

Fluency Practice (10 minutes)

- Take Out the Unit **5.NBT.1**   (4 minutes)
- Add and Subtract Decimals **5.NBT.7** (6 minutes)

Take Out the Unit (4 minutes)

Materials: (S) Personal white boards

Note: Decomposing common units as decimals will strengthen student understanding of place value.

- T: (Project 1.234 = _____ thousandths.) Say the number. Think about the how many thousandths in 1.234.
- T: (Project 1.234 = 1234 thousandths.) How much is one thousand, thousandths?
- S: One thousand, thousandths is the same as 1.
- T: (Project 65.247 = _____.) Say the number.
- S: 65 ones 247 thousandths.
- T: (Write 76.358 = 7 tens _____ thousandths.) On your board, fill in the blank.
- S: (Students write 76.358 = 7 tens 6358 thousandths.)

Repeat the process for hundredths 76.358 = 736 tenths _____ thousandths, 76.358 = _____ hundredths 8 thousandths.

Lesson 11: Multiply a decimal fraction by single-digit whole numbers, relate to a written method through application of the area model and place value understanding, and explain the reasoning used.
Date: 6/28/13

1.E.3

© 2013 Common Core, Inc. All rights reserved. commoncore.org

Add and Subtract Decimals (6 minutes)

Materials: (S) Personal white boards

Note: Reviewing these skills that were introduced in Lessons 9 and 10 will help students work towards mastery of adding and subtracting common decimal units.

T: (Write 7258 thousandths + 1 thousandth = ____.) Write the addition sentence in decimal form.

S: 7.258 + 0.001 = 7.259.

Repeat the process for 7 ones 258 thousandths + 3 hundredths, 7 ones 258 thousandths + 4 tenths, 6 ones 453 thousandths + 4 hundredths, 2 ones 37 thousandths + 5 tenths, and 6 ones 35 hundredths + 7 thousandths.

T: (Write 4 ones 8 hundredths – 2 ones = ___ ones ___ hundredths.) Write the subtraction sentence in decimal form.

S: (Students write 4.08 – 2 = 2.08.)

Repeat the process for 9 tenths 7 thousandths – 4 thousandths, 4 ones 582 thousandths – 3 hundredths, 9 ones 708 thousandths – 4 tenths, and 4 ones 73 thousandths – 4 hundredths.

Application Problems (5 minutes)

After school, Marcus ran 3.2km and Cindy ran 1.95km. Who ran farther? How much farther?

Note: This application problem requires students to subtract decimal numbers as studied in Lesson 10.

Concept Development (35 minutes)

Materials: (S) Personal white boards with place value charts, number disks

Problems 1–3

3 x 0.2 = 0.6
3 x 0.3 = 0.9
4 x 0.3 = 1.2

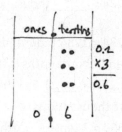

T: Place disks to show 2 tenths on your place value chart.

S: (Students draw.)

T: Make 3 copies of 2 tenths using number disks. How many tenths do you have in all?

S: Six tenths.

T: Turn to your partner and write a number sentence to express how we made 6 tenths.

S: I wrote 0.2 + 0.2 + 0.2 = 0.6 because I added 2 tenths

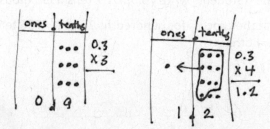

Lesson 11: Multiply a decimal fraction by single-digit whole numbers, relate to a written method through application of the area model and place value understanding, and explain the reasoning used.

Date: 6/28/13

COMMON CORE MATHEMATICS CURRICULUM • NY   Lesson 11   5•1

three times to get 6 tenths. → I multiplied 2 tenths by 3 and got 6 tenths so I wrote 3 x 0.2 = 0.6.

T: (Write on the board.) Complete the sentence: 3 copies of 2 tenths is _____; and read the equation in unit form: 3 x 0.2 = 0.6.

S: 6 tenths; 3 x 2 tenths = 6 tenths.

T: Work with your partner to find the value of 3 x 0.3 and 4 x 0.3.

S: (Students work and solve.)

T: How was 4 x 3 tenths different from 3 x 3 tenths?

S: I had to bundle the 10 tenths and made 1 one and had 2 tenths left, which I didn't do before. → We made a number greater than 1 whole.

T: 4 copies of 3 tenths is 12 tenths. (Show on place value chart.) 12 tenths is the same as _____.

S: 1 one and 2 tenths.

NOTES ON MULTIPLE MEANS OF ACTION AND EXPRESSION:

The area model can be considered a graphic organizer. It organizes the partial products. Some students may need support in order to remember which product goes in each cell of the area model especially as the model becomes more complex. Teachers can modify the organizer by writing the expressions in each cell. This might eliminate the need for some students to visually track the product into the appropriate cell.

Problems 4–6

2 x 0.43 = 0.86

2 x 0.423 = 0.846

4 x 0.423 = 1.692

T: (Write on chart.) 2 x 0.43 = _____. How can we use our knowledge from the previous problems to solve this?

S: We make copies of hundredths like we make copies of tenths. → Hundredths is a different unit, but we can multiply it just like tenths.

T: Use your place value chart to find the product of 2 x 0.43. Complete the sentence, "2 copies of 43 hundredths is _____."

S: (Students work.)

T: Read what your place value chart shows.

S: I have 2 groups of 4 tenths and 2 groups of 3 hundredths. I need to combine tenths with tenths and hundredths with hundredths.

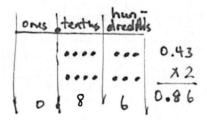
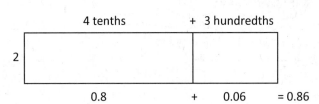

Lesson 11: Multiply a decimal fraction by single-digit whole numbers, relate to a written method through application of the area model and place value understanding, and explain the reasoning used.

Date: 6/28/13

T: (Teacher draws an area model.) Let me record what I hear you saying. Discuss with your partner the difference between these two models.

S: (Share observations.)

T: (Write on board.) 2 x 0.423 = _____. What is different about this problem?

S: There is a digit in the thousandths place. → We are multiplying thousandths.

T: Use your mat to solve this problem.

S: (Students work.)

T: Read what your place value chart shows.

S: 846 thousandths.

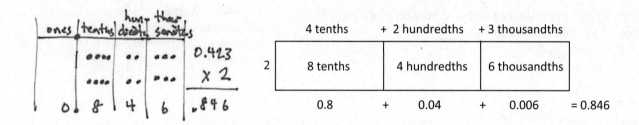

T: Now, draw an area model and write an equation with the partial products to show how you found the product.

S: (Students draw.)

T: (Write 4 x 0.423 = _____ on board.) Solve this with your disks.

S: (Students solve.)

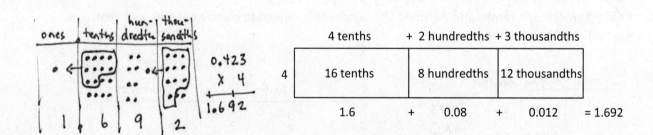

Lesson 11: Multiply a decimal fraction by single-digit whole numbers, relate to a written method through application of the area model and place value understanding, and explain the reasoning used.

Date: 6/28/13

T: Read the number that is shown on your chart.
S: 1 and 692 thousandths.
T: How was this problem different from the last?
S: 4 times 3 thousandths is 12 thousandths, so we had to bundle 10 thousandths to make 1 hundredth.
T: Did any other units have to be regrouped?
S: In the tenths place. Four times 4 tenths is 16 tenths, so we had to regroup 10 tenths to make 1 whole.
T: Let's record what happened on our mat using an area model and an equation showing the partial products.

> **NOTES ON MULTIPLE MEANS OF ENGAGEMENT:**
>
> It can be highly motivating for students to recognize their progress. Teachers can help students do this by creating a list of skills and concepts the students will master in this module. The students can keep track as the module and their skills progress.

Problems 7–9

(Use area model to represent distributive property.)

6 x 1.21

7 x 2.41

8 x 2.34

T: (Write on board.) 6 x 1.21. Let's imagine our disks, but use an area model to represent our thinking as we find the product of 6 times 1 and 21 hundredths.
T: (Draw area model on board.) On our area model, how many sections do we have?
S: 3. We have one for each place.
T: (Draw area model.) I have a section for 1 whole, 2 tenths, and 1 hundredth. I am multiplying each by what number?
S: 6.
T: With a partner, solve the equation using the area model and an equation which shows the partial products.
S: (Students work with a partner.)

Have students solve the last two equations using area models and recording equations. Circulate looking for any misconceptions.

Problem Set (10 minutes)

Students should do their personal best to complete the problem set within the allotted 10 minutes. For some classes, it may be appropriate to modify the assignment by specifying which problems they work on first. Some problems do not specify a method for solving. Students solve these problems using the RDW approach used for Application Problems.

Student Debrief (10 minutes)

Lesson Objective: Multiply a decimal fraction by single-digit whole numbers, relate to a written method through application of the area model and place value understanding, and explain the reasoning used.

The Student Debrief is intended to invite reflection and active processing of the total lesson experience.

Invite students to review their solutions for the Problem Set. They should check work by comparing answers with a partner before going over answers as a class. Look for misconceptions or misunderstandings that can be addressed in the Debrief. Guide students in a conversation to debrief the Problem Set and process the lesson. You may choose to use any combination of the questions below to lead the discussion.

- Compare student work in Problems 1(c) and 1(d) as some students may regroup units while others may not. Give opportunity for students to discuss the equality of the various unit decompositions. Give other examples (e.g., 6 x 0.25) asking students to defend the equality of 1.50, 150 hundredths, and 1.5 with words, models, and numbers.

- Problem 3 points out a common error in student thinking when multiplying decimals by whole numbers. Allow students to share their models for correcting Miles' error. Students should be able to articulate which units are being multiplied and composed into larger ones.

- Problem 3 also offers an opportunity to extend understanding by asking students to generate an area model and/or an equation using 6 as a multiplier that would make Miles' answer correct.

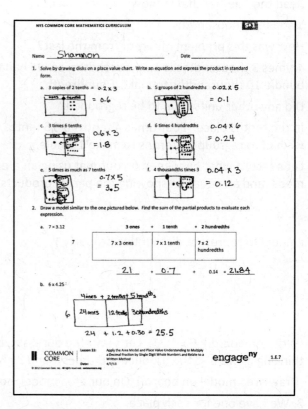

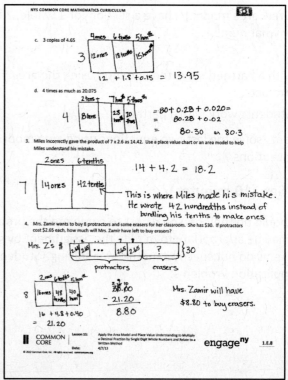

Exit Ticket (3 minutes)

After the Student Debrief, instruct students to complete the Exit Ticket. A review of their work will help you assess the students' understanding of the concepts that were presented in the lesson today and plan more effectively for future lessons. You may read the questions aloud to the students.

Lesson 11: Multiply a decimal fraction by single-digit whole numbers, relate to a written method through application of the area model and place value understanding, and explain the reasoning used.

Date: 6/28/13

Name _____   Date _____

1. Solve by drawing disks on a place value chart. Write an equation and express the product in standard form.

   a. 3 copies of 2 tenths

   b. 5 groups of 2 hundredths

   c. 3 times 6 tenths

   d. 6 times 4 hundredths

   e. 5 times as much as 7 tenths

   f. 4 thousandths times 3

2. Draw a model similar to the one pictured below for Parts (b), (c), and (d). Find the sum of the partial products to evaluate each expression.

   a. 7×3.12

	3 ones	**1 tenth**	**2 hundredths**
7	7 x 3 ones	7 x 1 tenth	7 x 2 hundredths

   _____ + _____ + 0.14 = _____

   b. 6×4.25

Lesson 11: Multiply a decimal fraction by single-digit whole numbers, relate to a written method through application of the area model and place value understanding, and explain the reasoning used.

Date: 6/28/13

c. 3 copies of 4.65

d. 4 times as much as 20.075

3. Miles incorrectly gave the product of 7 x 2.6 as 14.42. Use a place value chart or an area model to help Miles understand his mistake.

4. Mrs. Zamir wants to buy 8 protractors and some erasers for her classroom. She has $30. If protractors cost $2.65 each, how much will Mrs. Zamir have left to buy erasers?

COMMON CORE MATHEMATICS CURRICULUM • NY   Lesson 11 Exit Ticket   5•1

Name _____   Date _____

1. Solve by drawing disks on a place value chart. Write an equation and express the product in standard form.

   4 copies of 3 tenths

2. Complete the area model, and then find the product.

   3 × 9.63

   | 3 × ____ ones | 3 × ____ tenths | 4 × ____ hundredths |

Lesson 11: Multiply a decimal fraction by single-digit whole numbers, relate to a written method through application of the area model and place value understanding, and explain the reasoning used.

Date: 6/28/13

COMMON CORE MATHEMATICS CURRICULUM • NY   Lesson 11 Homework   5•1

Name _____   Date _____

1. Solve by drawing disks on a place value chart. Write an equation and express the product in standard form.

    a. 2 copies of 4 tenths

    b. 4 groups of 5 hundredths

    b. 4 times 7 tenths

    d. 3 times 5 hundredths

    c. 9 times as much as 7 tenths

    f. 6 thousandths times 8

2. Draw a model similar to the one pictured below. Find the sum of the partial products to evaluate each expression.

    a. 4 × 6.79

	6 ones +	**7 tenths** +	**9 hundredths**
4	4 × 6 ones	4 × 7 tenths	4 × 9 hundredths

    _____ + _____ + _____ = _____

Lesson 11: Multiply a decimal fraction by single-digit whole numbers, relate to a written method through application of the area model and place value understanding, and explain the reasoning used.
Date: 6/28/13

b. 6 x 7.49 hundredths

c. 9 copies of 3.65

d. 3 times 20.175

3. Leanne multiplied 8 x 4.3 and got 32.24. Is Leanne correct? Use an area model to explain your answer.

4. Anna buys groceries for her family. Hamburger meat is $3.38 per pound, sweet potatoes are $0.79 each, and hamburger rolls are $2.30 a bag. If Anna buys 3 pounds of meat, 5 sweet potatoes, and one bag of hamburger rolls, what will she pay in all for the groceries?

Lesson 12

Objective: Multiply a decimal fraction by single-digit whole numbers, including using estimation to confirm the placement of the decimal point.

Suggested Lesson Structure

- Fluency Practice   (12 minutes)
- Application Problems (8 minutes)
- Concept Development (30 minutes)
- Student Debrief   (10 minutes)
- **Total Time**   **(60 minutes)**

Fluency Practice (12 minutes)

- Add Decimals **5.NBT.7**   (9 minutes)
- Find the Product **5.NBT.7** (3 minutes)

Sprint: Add Decimals (9 minutes)

Materials: (S) Add Decimals Sprint

Note: This Sprint will help students build automaticity in adding decimals without renaming.

Find the Product (3 minutes)

Materials: (S) Personal white boards

Note: Reviewing this skill that was introduced in Lesson 11 will help students work towards mastery of multiplying single-digit numbers times decimals.

- T: (Write 4 x 2 ones = ___.) Write the multiplication sentence.
- S: 4 x 2 = 8
- T: Say the multiplication sentence in unit form.
- S: 4 x 2 ones = 8 ones.

Repeat the process for 4 x 0.2; 4 x 0.02; 5 x 3; 5 x 0.3; 5 x 0.03; 3 x 0.2; 3 x 0.03; 3 x 0.23; and 2 x 0.14.

COMMON CORE MATHEMATICS CURRICULUM • NY   Lesson 12   5•1

Application Problem (8 minutes)

Patty buys 7 juice boxes a month for lunch. If one juice costs $2.79, how much money does Patty spend on juice each month? Use an area model to solve.

Extension: How much will Patty spend on juice in 10 months? In 12 months?

```
Patty's $ [2.79][2.79]...[2.79]
          1box 2boxes ... 7boxes

           2    7tenths 9hun.
        ┌────┬────────┬─────┐
      7 │ 14 │  49    │ 63  │
        │    │ tenths │ hun.│
        └────┴────────┴─────┘
        14 + 4.9 + 0.63
        = $19.53
Patty spends $19.53 a month.
```

```
Bonus:
  19.53 × 10 = $195.30
              in 10 months

   19.53
  +19.53
  ─────
  $39.06 in 2 mths

12 mths:
  $195.30 + $39.06
  = $234.36 in 12 mths.
```

Note: The first part of this application problem asks students to multiply a number with two decimal digits by a single-digit whole number. This skill was taught in Module 1, Lesson 11 and provides a bridge to today's topic which involves reasoning about such problems on a more abstract level. The extension problem looks back to Topic A of this module, which requires multiplication by powers of 10. Students have not multiplied a decimal number by a two-digit number, but they are able to solve $2.79 × 12 by using the distributive property: 2.79 x (10 + 2).

Concept Development (30 minutes)

Materials: (S) Personal white boards

Problems 1–3

MP.8
31 x 4 = 124
3.1 x 4 = 12.4
0.31 x 4 = 0.124

T: (Write all 3 problems on board). How are these 3 problems alike?
S: They are alike because they all have 3, 1, and 4 as part of the problem.
T: Use an area model to find the products.
S: (Students draw.)

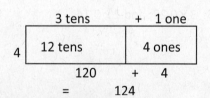
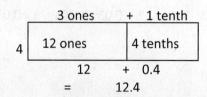
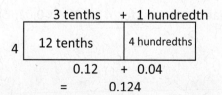

Lesson 12: Multiply a decimal fraction by single-digit whole numbers, including using estimation to confirm the placement of the decimal point.
Date: 6/28/13

T: How are the products of all three problems alike?

S: Every product has the digits 1, 2, and 4 and they are always in the same order.

T: If the products have the same digits and those digits are in the same order, do the products have the same value? Why or why not? Turn and talk.

S: No, the values are different because the units that we multiplied are different. → The decimal is not in the same place in every product. → The digits that we multiplied are the same, but you have to think about the units to make sure the answer is right.

T: So, let me repeat what I hear you saying. I can multiply the numerals first, then think about the units to help place the decimal.

MULTIPLE MEANS OF ACTION AND EXPRESSION:

Web based applications like Number Navigator offer assistance to those whose fine motor skills may prevent them from being able to set out columnar arithmetic with ease. Such applications preclude the need for complicated spreadsheets making them an ideal scaffold for the classroom.

Problems 4–6

5.1 x 6 = 30.6

11.4 x 4 = 45.6

7.8 x 3 = 23.4

T: (Write 5.1 x 6 on the board.) What is the smallest unit in 5.1?

S: Tenths.

T: Multiply 5.1 by 10 to convert it to tenths. How many tenths is the same as 5.1?

S: 51 tenths.

T: Suppose our multiplication sentence was 51 x 6. Multiply and record your multiplication vertically. What is the product?

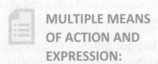

S: 306

T: We know that our product will contain these digits, but is 306 a reasonable product for our actual problem of 5.1 x 6? Turn and talk.

S: We have to think about the units. 306 ones is not reasonable, but 306 tenths is. → 5.1 is close to 5, and 5 x 6 = 30, so the answer should be around 30. → 306 tenths is the same as 30 ones and 6 tenths.

T: Using this reasoning, where does it make sense to place the decimal in 306? What is the product of 5.1 x 6?

S: Between the zero and the six. The product is 30.6.

T: (Write 11.4 x 4 = _____ on the board.) What is the smallest unit in 11.4?

S: Tenths.

T: What power of 10 must I use to convert 11.4 to tenths? How many tenths are the same as 11 ones 4 tenths? Turn and talk.

S: 10^1 → We have to multiply by 10. → 11.4 is the same as 114 tenths.

T: Multiply vertically to find the product of 114 tenths x 4.

S: 456 tenths.

T: We know that our product will contain these digits. How will we determine where to place our decimal?

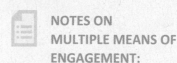

S: We can estimate. 11.4 is close to 11, and 11 x 4 is 44. The only place that makes sense for the decimal is between the five and six. The actual product is 45.6. → 456 tenths is the same as 45 ones and 6 tenths.

Repeat sequence with 7.8 x 3. Elicit from students the similarities and differences between this problem and others (must compose tenths into ones).

Problems 7–9

3.12 x 4 = 12.48

3.22 x 5 = 16.10

3.42 x 6 = 20.52

T: (Write 3.12 x 4 on board.) Use hundredths to name 3.12 and multiply vertically by 4. What is the product?

S: 1248 hundredths.

T: I will write 4 possible products for 3.12 x 4 on my board. Turn and talk to your partner about which of these products is reasonable. Then confirm the actual product using an area model. Be prepared to share your thinking. (Write 1248; 1.248; 12.48; 124.8 on board.)

S: (Students work and share.)

> **NOTES ON MULTIPLE MEANS OF ENGAGEMENT:**
>
> Once students are able to determine the reasonable placement of decimals through estimation, by composition of smaller units to larger units, and by using the area model, teachers should have students articulate which strategy they might choose first. Students who have choices develop self-determination and feel more connected to their learning.

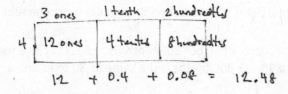

Repeat this sequence for the other problems in this set. Write possible products and allow students to reason about decimal placement both from an estimation-based strategy and from a composition of smaller units into larger units (i.e., 2,052 hundredths is the same as 20 ones and 52 hundredths). Students should also find the products using an area model and compare the two methods for finding products.

Lesson 12: Multiply a decimal fraction by single-digit whole numbers, including using estimation to confirm the placement of the decimal point.

Date: 6/28/13

Problems 10–12

0.733 × 4 = 2.932

10.733 × 4 = 42.932

5.733 × 4 = 22.932

> T: (Write 0.733 × 4 on board.) Rename 0.733 using its smallest units and multiply vertically by 4. What is the product?
>
> S: 2932 thousandths.
>
> T: (Write 2.932; 29.32; 293.2; and 2,932 on board.) Which of these is the most reasonable product for 0.733 × 4? Why? Turn and talk.
>
> S: 2.932, because 0.733 is close to one whole and 1 × 4 = 4. None of the other choices make sense. → I know that 2000 thousandths make 2 wholes, so 2932 thousandths is the same as 2 ones 932 thousandths.
>
> T: Solve 0.733 × 4 using an area model. Compare your products using these two different strategies.
>
> S: (Students work.)

Repeat this sequence for 10.733 × 4 and allow independent work for 5.733 × 4. Require students to use decomposition to smallest units, reason about decimal placement and the area model so that products and strategies may be compared.

Problem Set (10 minutes)

Students should do their personal best to complete the Problem Set within the allotted 10 minutes. For some classes, it may be appropriate to modify the assignment by specifying which problems they work on first. Some problems do not specify a method for solving. Students solve these problems using the RDW approach used for Application Problems.

Student Debrief (10 minutes)

Lesson Objective: Multiply a decimal fraction by single-digit whole numbers, including using estimation to confirm the placement of the decimal point

The Student Debrief is intended to invite reflection and active processing of the total lesson experience.

Invite students to review their solutions for the Problem Set. They should check work by comparing answers with a partner before going over answers as a class. Look for misconceptions or misunderstandings that can be addressed in the Debrief. Guide students in a

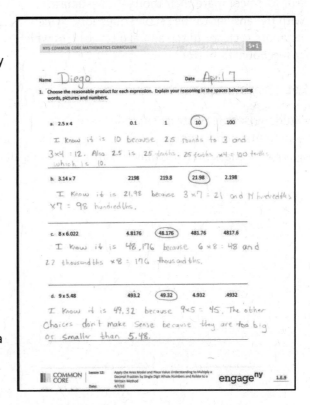

Lesson 12: Multiply a decimal fraction by single-digit whole numbers, including using estimation to confirm the placement of the decimal point.

Date: 6/28/13

COMMON CORE MATHEMATICS CURRICULUM • NY   Lesson 12   5•1

conversation to debrief the Problem Set and process the
lesson. You may choose to use any combination of the
questions below to lead the discussion.

- How can whole number multiplication help you
  with decimal multiplication? (Elicit from students
  that the digits in a product can be found through
  whole number multiplication. The actual product
  can be deduced through estimation based logic
  and/or composing smaller units into larger units.)
- How does the area model help you to justify the
  placement of the decimal point for the product in
  1(b)?
- Problem 3 offers an excellent opportunity to
  discuss purposes of estimation because multiple
  answers are possible for the estimate Marcel
  gives his gym teacher. (For example, do we
  round to 4 and estimate that he bikes about 16
  miles? Or do we round to 3.5 because out and
  back gives us 7 miles each time, which is 14 miles
  altogether?) Allow time for students to debate
  the thinking behind their choices. It may also be
  fruitful to compare their thoughtful estimates
  with the answer to the second question. Which
  estimate is closer to the actual distance? In which
  cases would it matter?

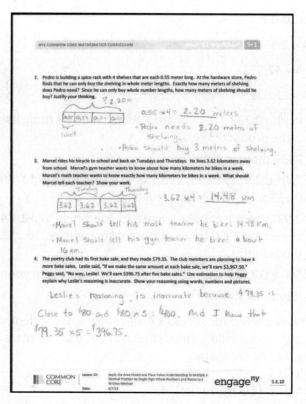

Exit Ticket (3 minutes)

After the Student Debrief, instruct students to complete the Exit Ticket. A review of their work will help you assess the students' understanding of the concepts that were presented in the lesson today and plan more effectively for future lessons. You may read the questions aloud to the students.

Lesson 12: Multiply a decimal fraction by single-digit whole numbers, including
             using estimation to confirm the placement of the decimal point.
Date: 6/28/13

A

Add.

Correct _____

#			#		
1	3 + 1 =		23	5 + 0.1 =	
2	3.5 + 1 =		24	5.7 + 0.1 =	
3	3.52 + 1 =		25	5.73 + 0.1 =	
4	0.3 + 0.1 =		26	5.736 + 0.1 =	
5	0.37 + 0.1 =		27	5.736 + 1 =	
6	5.37 + 0.1 =		28	5.736 + 0.01 =	
7	0.03 + 0.01 =		29	5.736 + 0.001 =	
8	0.83 + 0.01 =		30	6.208 + 0.01 =	
9	2.83 + 0.01 =		31	3 + 0.01 =	
10	30 + 10 =		32	3.5 + 0.01 =	
11	32 + 10 =		33	3.58 + 0.01 =	
12	32.5 + 10 =		34	3.584 + 0.01 =	
13	32.58 + 10 =		35	3.584 + 0.001 =	
14	40.789 + 1 =		36	3.584 + 0.1 =	
15	4 + 1 =		37	3.584 + 1 =	
16	4.6 + 1 =		38	6.804 + 0.01 =	
17	4.62 + 1 =		39	8.642 + 0.001 =	
18	4.628 + 1 =		40	7.65 + 0.001 =	
19	4.628 + 0.1 =		41	3.987 + 0.1 =	
20	4.628 + 0.01 =		42	4.279 + 0.001 =	
21	4.628 + 0.001 =		43	13.579 + 0.01 =	
22	27.048 + 0.1 =		44	15.491 + 0.01 =	

© Bill Davidson

Lesson 12: Multiply a decimal fraction by single-digit whole numbers, including using estimation to confirm the placement of the decimal point.
Date: 6/28/13

B   Add.   Improvement _____   # Correct _____

#	Problem		#	Problem	
1	2 + 1 =		23	4 + 0.1 =	
2	2.5 + 1 =		24	4.7 + 0.1 =	
3	2.53 + 1 =		25	4.73 + 0.1 =	
4	0.2 + 0.1 =		26	4.736 + 0.1 =	
5	0.27 + 0.1 =		27	4.736 + 1 =	
6	5.27 + 0.1 =		28	4.736 + 0.01 =	
7	0.02 + 0.01 =		29	4.736 + 0.001 =	
8	0.82 + 0.01 =		30	5.208 + 0.01 =	
9	4.82 + 0.01 =		31	2 + 0.01 =	
10	20 + 10 =		32	2.5 + 0.01 =	
11	23 + 10 =		33	2.58 + 0.01 =	
12	23.5 + 10 =		34	2.584 + 0.01 =	
13	23.58 + 10 =		35	2.584 + 0.001 =	
14	30.789 + 1 =		36	2.584 + 0.1 =	
15	3 + 1 =		37	2.584 + 1 =	
16	3.6 + 1 =		38	5.804 + 0.01 =	
17	3.62 + 1 =		39	7.642 + 0.001 =	
18	3.628 + 1 =		40	6.75 + 0.001 =	
19	3.628 + 0.1 =		41	2.987 + 0.1 =	
20	3.628 + 0.01 =		42	3.279 + 0.001 =	
21	3.628 + 0.001 =		43	12.579 + 0.01 =	
22	37.048 + 0.1 =		44	14.391 + 0.01 =	

© Bill Davidson

Lesson 12: Multiply a decimal fraction by single-digit whole numbers, including using estimation to confirm the placement of the decimal point.
Date: 6/28/13

COMMON CORE MATHEMATICS CURRICULUM • NY   Lesson 12 Problem Set   5•1

Name _____   Date _____

1. Choose the reasonable product for each expression. Explain your reasoning in the spaces below using words, pictures and numbers.

   a. 2.5 x 4   0.1   1   10   100

   b. 3.14 x 7   2198   219.8   21.98   2.198

   c. 8 x 6.022   4.8176   48.176   481.76   4817.6

   d. 9 x 5.48   493.2   49.32   4.932   .4932

Lesson 12: Multiply a decimal fraction by single-digit whole numbers, including using estimation to confirm the placement of the decimal point.
Date: 6/28/13

1.E.23

2. Pedro is building a spice rack with 4 shelves that are each 0.55 meter long. At the hardware store, Pedro finds that he can only buy the shelving in whole meter lengths. Exactly how many meters of shelving does Pedro need? Since he can only buy whole number lengths, how many meters of shelving should he buy? Justify your thinking.

3. Marcel rides his bicycle to school and back on Tuesdays and Thursdays. He lives 3.62 kilometers away from school. Marcel's gym teacher wants to know about how many kilometers he bikes in a week. Marcel's math teacher wants to know exactly how many kilometers he bikes in a week. What should Marcel tell each teacher? Show your work.

4. The poetry club had its first bake sale, and they made $79.35. The club members are planning to have 4 more bake sales. Leslie said, "If we make the same amount at each bake sale, we'll earn $3,967.50." Peggy said, "No way, Leslie! We'll earn $396.75 after five bake sales." Use estimation to help Peggy explain why Leslie's reasoning is inaccurate. Show your reasoning using words, numbers and pictures.

COMMON CORE MATHEMATICS CURRICULUM • NY   Lesson 12 Exit Ticket   5•1

Name _____   Date _____

1. Use estimation to choose the correct value for each expression.

   a. 5.1 x 2   0.102   1.02   10.2   102

   b. 4 x 8.93   3.572   35.72   357.2   3572

2. Estimate the answer for 7.13 x 6. Explain your reasoning using words, pictures or numbers.

COMMON CORE MATHEMATICS CURRICULUM • NY   Lesson 12 Homework   5•1

Name _____   Date _____

1. Choose the reasonable product for each expression. Explain your thinking in the spaces below using words, pictures, and numbers.

   a.   2.1 x 3   0.63   6.3   63   630

   _____

   b.   4.27 x 6   2562   256.2   25.62   2.562

   _____

   c.   7 x 6.053   4237.1   423.71   42.371   4.2371

   _____

   d.   9 x 4.82   4.338   43.38   433.8   4338

   _____

2. YiTing weighs 8.3 kg. Her older brother is 4 times as heavy as her. How much does her older brother's weight in kg?

Lesson 12: Multiply a decimal fraction by single-digit whole numbers, including using estimation to confirm the placement of the decimal point.
Date: 6/28/13

3. Tim is painting his storage shed. He buys 4 gallons of white paint and 3 gallons of blue paint. If each gallon of white paint costs $15.72 and each gallon of blue paints is $21.87, how much will Tim spend in all on paint?

4. Ribbon is sold at 3 yards for $6.33. Jackie bought 24 yards of ribbon for a project. How much did she pay?

COMMON CORE MATHEMATICS CURRICULUM • NY　　　　　　　　　　　　　Topic F　5•1

GRADE 5 • MODULE 1

Topic F
Dividing Decimals

5.NBT.3, 5.NBT.7

Focus Standard:	5.NBT.3	Read, write, and compare decimals to thousandths. a. Read and write decimals to thousandths using base-ten numerals, number names, and expanded form, e.g., 347.392 = 3 × 100 + 4 × 10 + 7 × 1 + 3 × (1/10) + 9 × (1/100) + 2 × (1/1000). b. Compare two decimals to thousandths based on meanings of the digits in each place, using >, =, and < symbols to record the results of comparisons.
	5.NBT.7	Add, subtract, multiply and divide decimals to hundredths, using concrete models or drawings and strategies based on place value, properties of operations, and/or the relationship between addition and subtraction; relate the strategy to a written method and explain the reasoning used.
Instructional Days:	4	
Coherence -Links from:	G4–M3	Multi-Digit Multiplication and Division
-Links to:	G5–M2	Multi-Digit Whole Number and Decimal Fraction Operations
	G6–M2	Arithmetic Operations Including Dividing by a Fraction

Topic F concludes Module 1 with an exploration of division of decimal numbers by one-digit whole number divisors using place value charts and disks. Lessons begin with easily identifiable multiples such as 4.2 ÷ 6 and move to quotients which have a remainder in the smallest unit (through the thousandths). Written methods for decimal cases are related to place value strategies, properties of operations and familiar written methods for whole numbers (**5.NBT.7**). Students solidify their skills with an understanding of the algorithm before moving on to division involving two-digit divisors in Module 2. Students apply their accumulated knowledge of decimal operations to solve word problems at the close of the module.

COMMON CORE MATHEMATICS CURRICULUM • NY

Topic F 5•1

A Teaching Sequence Towards Mastery of Dividing Decimals

Objective 1: Divide decimals by single-digit whole numbers involving easily identifiable multiples using place value understanding and relate to a written method.
(Lesson 13)

Objective 2: Divide decimals with a remainder using place value understanding and relate to a written method.
(Lesson 14)

Objective 3: Divide decimals using place value understanding including remainders in the smallest unit.
(Lesson 15)

Objective 4: Solve word problems using decimal operations.
(Lesson 16)

Lesson 13

Objective: Divide decimals by single-digit whole numbers involving easily identifiable multiples using place value understanding and relate to a written method.

Suggested Lesson Structure

- **Fluency Practice**   (15 minutes)
- **Application Problems**   (7 minutes)
- **Concept Development**   (28 minutes)
- **Student Debrief**   (10 minutes)
- **Total Time**   **(60 minutes)**

Fluency Practice (15 minutes)

- Subtract Decimals   **5.NBT.7**   (9 minutes)
- Find the Product   **5.NBT.7**   (3 minutes)
- Compare Decimal Fractions   **3.NF.3d**   (3 minutes)

Sprint: Subtract Decimals (9 minutes)

Materials: (S) Subtract Decimals Sprint

Note: This Sprint will help students build automaticity in subtracting decimals without renaming.

Find the Product (3 minutes)

Materials: (S) Personal white boards

Note: Reviewing this skill that was introduced in Lessons 11 and 12 will help students work towards mastery of multiplying single-digit numbers times decimals.

- T: (Write 4 x 3 = ___.) Say the multiplication sentence in unit form.
- S: 4 x 3 ones = 12 ones.
- T: (Write 4 x 0.2 = ___.) Say the multiplication sentence in unit form.
- S: 4 x 2 tenths = 8 tenths.
- T: (Write 4 x 3.2 = ___.) Say the multiplication sentence in unit form.
- S: 4 x 3 ones 2 tenths = 12.8.
- T: Write the multiplication sentence.

COMMON CORE MATHEMATICS CURRICULUM • NY   Lesson 13 5•1

S: (Students write 4 x 3.1 = 12.8.)

Repeat the process for 4 x 3.21, 9 x 2, 9 x 0.1, 9 x 0.03, 9 x 2.13, 4.012 x 4, and 5 x 3.2375.

Compare Decimal Fractions (3 minutes)

Materials: (S) Personal white boards

Note: This review fluency will help solidify student understanding of place value in the decimal system.

T: (Write 13.78 ___ 13.86.) On your personal white boards, compare the numbers using the greater than, less than, or equal sign.

S: (Students write 13.78 < 13.76.)

Repeat the process and procedure for 0.78 ___ 78/100, 439.3 ___ 4.39, 5.08 ___ fifty-eight tenths, Thirty-five and 9 thousandths ___ 4 tens.

Application Problems (7 minutes)

Louis buys 4 chocolates. Each chocolate costs $2.35. Louis multiplies 4 x 235 and gets 940. Place the decimal to show the cost of the chocolates and explain your reasoning using words, numbers, and pictures.

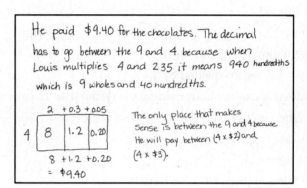

Note: This application problem requires students to estimate 4 × $2.35 in order to place the decimal point in the product. This skill was taught in the previous lesson.

Concept Development (28 minutes)

Materials: (S) Number disks, personal white boards

Problems 1–3

0.9 ÷ 3 = 0.3

0.24 ÷ 4 = 0.06

0.032 ÷ 8 = 0.004

T: Show 9 tenths with your disks.
S: (Students show.)
T: Divide 9 tenths into 3 equal groups.
S: (Students make 3 groups of 3 tenths.)
T: How many tenths are in each group?
S: There are 3 tenths in each group.

Lesson 13: Divide decimals by single-digit whole numbers involving easily identifiable multiples using place value understanding and relate to a written method.

Date: 6/28/13

1.F.4

T: (Write 0.9 ÷ 3 = 0.3 on board.) Read the number sentence using unit form.
S: 9 tenths divided by 3 equals 3 tenths.
T: How does unit form help us divide?
S: When we identify the units, then it's just like dividing 9 apples into 3 groups. → If you know what unit you are sharing, then it's just like whole number division. You can just think about the basic fact.
T: (Write 3 groups of _____ = 0.9 on board.) What is the missing number in our equation?
S: 3 tenths (0.3).

Repeat this sequence with 0.24 (24 hundredths) and 0.032 (32 thousandths).

Problems 4–6

1.5 ÷ 5 = 0.3

1.05 ÷ 5 = 0.21

3.015 ÷ 5 = 0.603

T: (Write on board.) 1.5 ÷ 5 = _____. Read the equation using unit form.
S: Fifteen tenths divided by 5.
T: What is useful about reading the decimal as 15 tenths?
S: When you say the units, it's like a basic fact.
T: What is 15 tenths divided by 5?
S: 3 tenths.
T: (Write on board.) 1.5 ÷ 5 = 0.3
T: (Write on board.) 1.05 ÷ 5 = _____. Read the equation using unit form.
S: 105 hundredths divided by 5.
T: Is there another way to decompose (name or group) this quantity?
S: 1 one and 5 hundredths. → 10 tenths and 5 hundredths.
T: Which way of naming 1.05 is most useful when dividing by 5? Why? Turn and talk. Then solve.
S: 10 tenths and 5 hundredths because they are both multiples of 5. This makes it easy to use basic facts and divide mentally. The answer is 2 tenths and 1 hundredth. → 105 hundredths is easier for me because I know 100 is 20 fives so 105 is 1 more, 21. 21 hundredths. → I just used the algorithm from Grade 4 and got 21 and knew it was hundredths.

Repeat this sequence with 3.015 ÷ 5. Have students decompose the decimal several ways and then reason about which is the most useful for division. It is also important to draw parallels among the next three problems. You might ask, "How does the answer to the second set of problems help you find the answer to the third?"

NOTES ON MULTIPLE MEANS OF ENGAGEMENT:

Students can also be challenged to use a compensation strategy to make another connection to whole number division. The dividend is multiplied by a power of ten, which converts it to its smallest units. Once the dividend is shared among the groups, it must be converted back to the original units by dividing it by the same power of ten. For example :

1.5 ÷ 5 → (1.5 **x 10**) ÷ 5 →

15 ÷ 5 = 3 → 3 ÷ **10** = 0.3

Lesson 13: Divide decimals by single-digit whole numbers involving easily identifiable multiples using place value understanding and relate to a written method.
Date: 6/28/13

COMMON CORE MATHEMATICS CURRICULUM • NY   Lesson 13   5•1

Problems 7–9

Compare the relationships between:

4.8 ÷ 6 = 0.8 and 48 ÷ 6 = 8

4.08 ÷ 8 = 0.51 and 408 ÷ 8 = 51

63.021 ÷ 7 = 9.003 and 63,021 ÷ 7 = 9,003

T: (Write on board 4.8 ÷ 6 = 0.8   48 ÷ 6 = 8.) What relationships do you notice between these two equations? How are they alike?

S: 8 is 10 times greater than 0.8. → 48 is 10 times greater than 4.8 → The digits in the dividends are the same, the divisor is the same and the digits in the quotient are the same.

T: How can 48 ÷ 6 help you with 4.8 ÷ 6? Turn and talk.

S: If you think of the basic fact first, then you can get a quick answer. Then you just have to remember what units were really in the problem. This one was really 48 tenths → The division is the same; the units are the only difference.

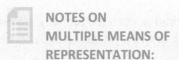

NOTES ON MULTIPLE MEANS OF REPRESENTATION:

Unfamiliar vocabulary can slow down the learning process, or even confuse students. Reviewing key vocabulary, such as dividend, divisor, or quotient may benefit all students. Displaying the words in a familiar mathematical sentence may serve as a useful reference for students. For example, display:

Dividend ÷ **Divisor** = Quotient.

Repeat the process for following equations:

4.08 ÷ 8 = 0.51 and 408 ÷ 8 = 51; 63.021 ÷ 7 = 9.003 and 63,021 ÷ 7 = 9,003

T: When completing your problem set, remember to use what you know about whole numbers to help you divide the decimals.

Problem Set (10 minutes)

Students should do their personal best to complete the problem set within the allotted 10 minutes. For some classes, it may be appropriate to modify the assignment by specifying which problems they work on first. Some problems do not specify a method for solving. Students solve these problems using the RDW approach used for Application Problems.

Student Debrief (10 minutes)

Lesson Objective: Divide decimals by single-digit whole numbers involving easily identifiable multiples using place value understanding and relate to a written method.

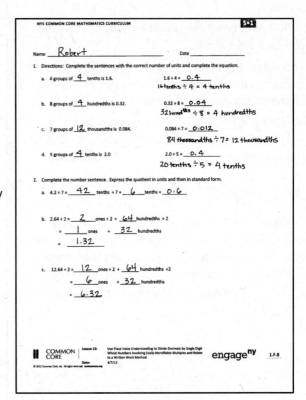

Lesson 13:   Divide decimals by single-digit whole numbers involving easily identifiable multiples using place value understanding and relate to a written method.
Date: 6/28/13

1.F.6

COMMON CORE MATHEMATICS CURRICULUM • NY   Lesson 13   5•1

The Student Debrief is intended to invite reflection and active processing of the total lesson experience.

Invite students to review their solutions for the Problem Set. They should check work by comparing answers with a partner before going over answers as a class. Look for misconceptions or misunderstandings that can be addressed in the Debrief. Guide students in a conversation to debrief the Problem Set and process the lesson. You may choose to use any combination of the questions below to lead the discussion.

- In 2(a), how does your understanding of whole number division help you solve the equation with a decimal?
- Is there another decomposition of the dividend in 2(c) that could have been useful in dividing by 2? What about in 2(d)? Why or why not?
- When decomposing decimals in different ways, how can you tell which is the most useful? (We are looking for easily identifiable multiples of the divisor.)
- In 4(a), what mistake is being made that would produce 5.6 ÷ 7 = 8?
- Correct all the dividends in Problem 4 so that the quotients are correct. Is there a pattern to the changes that you must make?
- 4.221 ÷ 7 = _____. Explain how you would decompose 4.221 so that you only need knowledge of basic facts to find the quotient.

Exit Ticket (3 minutes)

After the Student Debrief, instruct students to complete the Exit Ticket. A review of their work will help you assess the students' understanding of the concepts that were presented in the lesson today and plan more effectively for future lessons. You may read the questions aloud to the students.

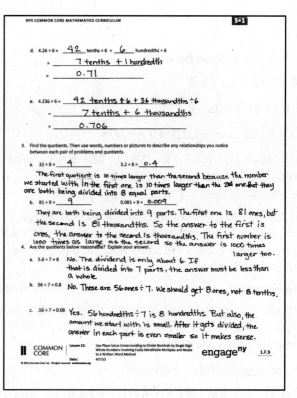

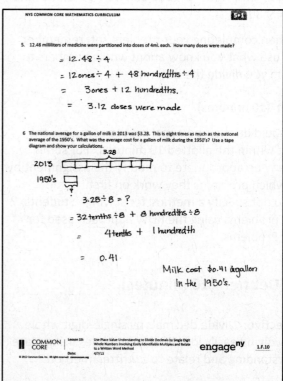

Lesson 13: Divide decimals by single-digit whole numbers involving easily identifiable multiples using place value understanding and relate to a written method.
Date: 6/28/13

A

Subtract.   # Correct _____

#	Problem		#	Problem	
1	5 - 1 =	.	23	7.985 - 0.002 =	.
2	5.9 - 1 =	.	24	7.985 - 0.004 =	.
3	5.93 - 1 =	.	25	2.7 - 0.1 =	.
4	5.932 - 1 =	.	26	2.785 - 0.1 =	.
5	5.932 - 2 =	.	27	2.785 - 0.5 =	.
6	5.932 - 4 =	.	28	4.913 - 0.4 =	.
7	0.5 - 0.1 =	.	29	3.58 - 0.01 =	.
8	0.53 - 0.1 =	.	30	3.586 - 0.01 =	.
9	0.539 - 0.1 =	.	31	3.586 - 0.05 =	.
10	8.539 - 0.1 =	.	32	7.982 - 0.04 =	.
11	8.539 - 0.2 =	.	33	6.126 - 0.001 =	.
12	8.539 - 0.4 =	.	34	6.126 - 0.004 =	.
13	0.05 - 0.01 =	.	35	9.348 - 0.006 =	.
14	0.057 - 0.01 =	.	36	8.347 - 0.3 =	.
15	1.057 - 0.01 =	.	37	9.157 - 0.05 =	.
16	1.857 - 0.01 =	.	38	6.879 - 0.009 =	.
17	1.857 - 0.02 =	.	39	6.548 - 2 =	.
18	1.857 - 0.04 =	.	40	6.548 - 0.2 =	.
19	0.005 - 0.001 =	.	41	6.548 - 0.02 =	.
20	7.005 - 0.001 =	.	42	6.548 - 0.002 =	.
21	7.905 - 0.001 =	.	43	6.196 - 0.06 =	.
22	7.985 - 0.001 =	.	44	9.517 - 0.004 =	.

© Bill Davidson

Lesson 13: Divide decimals by single-digit whole numbers involving easily identifiable multiples using place value understanding and relate to a written method.

Date: 6/28/13

1.F.8

B Subtract. Improvement _____   # Correct _____

#	Problem	#	Problem
1	6 - 1 =	23	7.986 - 0.002 =
2	6.9 - 1 =	24	7.986 - 0.004 =
3	6.93 - 1 =	25	3.7 - 0.1 =
4	6.932 - 1 =	26	3.785 - 0.1 =
5	6.932 - 2 =	27	3.785 - 0.5 =
6	6.932 - 4 =	28	5.924 - 0.4 =
7	0.6 - 0.1 =	29	4.58 - 0.01 =
8	0.63 - 0.1 =	30	4.586 - 0.01 =
9	0.639 - 0.1 =	31	4.586 - 0.05 =
10	8.639 - 0.1 =	32	6.183 - 0.04 =
11	8.639 - 0.2 =	33	7.127 - 0.001 =
12	8.639 - 0.4 =	34	7.127 - 0.004 =
13	0.06 - 0.01 =	35	1.459 - 0.006 =
14	0.067 - 0.01 =	36	8.457 - 0.4 =
15	1.067 - 0.01 =	37	1.267 - 0.06 =
16	1.867 - 0.01 =	38	7.981 - 0.001 =
17	1.867 - 0.02 =	39	7.548 - 2 =
18	1.867 - 0.04 =	40	7.548 - 0.2 =
19	0.006 - 0.001 =	41	7.548 - 0.02 =
20	7.006 - 0.001 =	42	7.548 - 0.002 =
21	7.906 - 0.001 =	43	7.197 - 0.06 =
22	7.986 - 0.001 =	44	1.627 - 0.004 =

© Bill Davidson

Lesson 13: Divide decimals by single-digit whole numbers involving easily identifiable multiples using place value understanding and relate to a written method.

Date: 6/28/13

COMMON CORE MATHEMATICS CURRICULUM • NY   Lesson 13 Problem Set   5•1

Name _____   Date _____

1. Complete the sentences with the correct number of units and complete the equation.

    a. 4 groups of _____ tenths is 1.6.   1.6 ÷ 4 = _____

    b. 8 groups of _____ hundredths is 0.32.   0.32 ÷ 8 = _____

    c. 7 groups of _____ thousandths is 0.084.   .084 ÷ 7 = _____

    d. 5 groups of _____ tenths is 2.0   2.0 ÷ 5 = _____

2. Complete the number sentence. Express the quotient in units and then in standard form.

    a. 4.2 ÷ 7 = _____ tenths ÷ 7 = _____ tenths = _____

    b. 2.64 ÷ 2 = _____ ones ÷ 2 + _____ hundredths ÷ 2

       = _____ ones + _____ hundredths

       = _____

    c. 12.64 ÷ 2 = _____ ones ÷ 2 + _____ hundredths ÷ 2

       = _____ ones + _____ hundredths

       = _____

Lesson 13: Divide decimals by single-digit whole numbers involving easily identifiable multiples using place value understanding and relate to a written method.
Date: 6/28/13

d. 4.26 ÷ 6 = _____ tenths ÷ 6 + _____ hundredths ÷ 6

= _____

= _____

e. 4.236 ÷ 6 = _____

= _____

= _____

3. Find the quotients. Then use words, numbers, or pictures to describe any relationships you notice between each pair of problems and quotients.

   a. 32 ÷ 8 = _____   3.2 ÷ 8 = _____

   b. 81 ÷ 9 = _____   0.081 ÷ 9 = _____

4. Are the quotients below reasonable? Explain your answer.

   a. 5.6 ÷ 7 = 8

   b. 56 ÷ 7 = 0.8

   c. .56 ÷ 7 = 0.08

5. 12.48 milliliters of medicine were separated into doses of 4 ml each. How many doses were made?

6. The price of most milk in 2013 is around $3.28 a gallon. This is eight times as much as you would have probably paid for a gallon of milk in the 1950's. What was the cost for a gallon of milk during the 1950's? Use a tape diagram and show your calculations.

Name _____   Date _____

1. Complete the sentences with the correct number of units and complete the equation.

    a. 2 groups of _____ tenths is 1.8   1.8 ÷ 2 = _____

    b. 4 groups of _____ hundredths is 0.32   0.32 ÷ 4 = _____

    c. 7 groups of _____ thousandths is 0.021   0.021 ÷ 7 = _____

2. Complete the number sentence. Express the quotient in units and then in standard form.

    a. 4.5 ÷ 5 = _____ tenths ÷ 5 = _____ tenths = _____

    b. 6.12 ÷ 6 = _____ ones ÷ 6 + _____ hundredths ÷ 6

         = _____ ones + _____ hundredths

         = _____

COMMON CORE MATHEMATICS CURRICULUM • NY   Lesson 13 Homework   5•1

Name _____   Date _____

1. Complete the sentences with the correct number of units and complete the equation.

   a. 3 groups of _____ tenths is 1.5 $1.5 \div 3 =$ _____

   b. 6 groups of _____ hundredths is 0.24 $0.24 \div 6 =$ _____

   c. 5 groups of _____ thousandths is 0.045 $0.045 \div 5 =$ _____

2. Complete the number sentence. Express the quotient in units and then in standard form.

   a. $9.36 \div 3 =$ _____ ones $\div$ 3 + _____ hundredths $\div$ 3

      = _____ ones + _____ hundredths

      = _____

   b. $36.012 \div 3 =$ _____ ones $\div$ 3 + _____ thousandths $\div$ 3

      = _____ ones + _____ thousandths

      = _____

   c. $3.55 \div 5 =$ _____ tenths $\div$ 5 + _____ hundredths $\div$ 5

      = _____

      = _____

Lesson 13: Divide decimals by single-digit whole numbers involving easily identifiable multiples using place value understanding and relate to a written method.
Date: 6/28/13

d. 3.545 ÷ 5 = _____

= _____

= _____

3. Find the quotients. Then use words, numbers, or pictures to describe any relationships you notice between each pair of problems and quotients.

   a. 21 ÷ 7 = _____   2.1 ÷ 7 = _____

   b. 48 ÷ 8 = _____   0.048 ÷ 8 = _____

4. Are the quotients below reasonable? Explain your answer.

   a. 0.54 ÷ 6 = 9

   b. 5.4 ÷ 6 = 0.9

c. 54 ÷ 6 = 0.09

5. A toy airplane costs $4.84. It costs 4 times as much as a toy car. What is the cost of the toy car?

6. Julian bought 3.9 liters of cranberry juice and Jay bought 8.74 liters of apple juice. They mixed the two juices together then poured them equally into 2 bottles. How many liters of juice are in each bottle?

Lesson 14

Objective: Divide decimals with a remainder using place value understanding and relate to a written method.

Suggested Lesson Structure

- Fluency Practice   (12 minutes)
- Application Problems (8 minutes)
- Concept Development   (30 minutes)
- Student Debrief   (10 minutes)
- **Total Time**   **(60 minutes)**

Fluency Practice (12 minutes)

- Multiply and Divide by Exponents **5.NBT.2**   (3 minutes)
- Round to Different Place Values **5.NBT.4**   (3 minutes)
- Find the Quotient **5.NBT.5**   (6 minutes)

Multiply and Divide by Exponents (3 minutes)

Materials: (S) Personal white boards

Notes: This review fluency will help solidify student understanding of multiplying by 10, 100, and 1000 in the decimal system.

- T: (Project place value chart from millions to thousandths.) Write 65 tenths as a decimal. Students write 6 in the ones column and 5 in the tenths column.
- T: Say the decimal.
- S: 6.5
- T: Multiply it by 10^2.
- S: (Students cross out 6.5 and write 650.)

Repeat the process and sequence for 0.7×10^2, $0.8 \div 10^2$, 3.895×10^3, and $5472 \div 10^3$

COMMON CORE MATHEMATICS CURRICULUM • NY   Lesson 14 5•1

Round to Different Place Values (3 minutes)

Materials: (S) Personal white boards

Notes: This review fluency will help solidify student understanding of rounding decimals to different place values.

- T: (Project 6.385.) Say the number.
- S: 6 and 385 thousandths.
- T: On your boards, round the number to the nearest tenth.
- S: (Students write 6.385 ≈ 6.4.)

Repeat the process, rounding 6.385 to the nearest hundredth. Follow the same process, but vary the sequence for 37.645.

Find the Quotient (6 minutes)

Materials: (S) Personal white boards

Notes: Reviewing these skills that were introduced in Lesson 13 will help students work towards mastery of dividing decimals by single-digit whole numbers.

- T: (Write 14 ÷ 2 = ___.) Write the division sentence.
- S: 14 ÷ 2 = 7.
- T: Say the division sentence in unit form.
- S: 14 ones ÷ 2 = 7 ones.

Repeat the process for 1.4 ÷ 2, 0.14 ÷ 2, 24 ÷ 3, 2.4 ÷ 3, 0.24 ÷ 3, 30 ÷ 3, 3 ÷ 5, 4 ÷ 5, and 2 ÷ 5.

Application Problems (8 minutes)

A bag of potato chips contains 0.96 grams of sodium. If the bag is split into 8 equal servings, how many grams of sodium will each serving contain?

Bonus: What other ways can the bag be divided into equal servings so that the amount of sodium in each serving has two digits to the right of the decimal and the digits are greater than zero in the tenths and hundredths place?

```
0.96 ÷ 8
= 96 hundredths ÷ 8
= 12 hundredths
= 0.12g of sodium per serving

Bonus:
96 can be divided by:
2 → 0.48g ✓        7 → too many decimal ✗
3 → 0.32g ✓             places
4 → 0.24g ✓        9 → too many decimal ✗
5 → 0.192g ✗            places
6 → 0.16g         10 → less than 0.11 ✗
```

Lesson 14: Divide decimals with a remainder using place value understanding and relate to a written method.
Date: 6/28/13

COMMON CORE MATHEMATICS CURRICULUM • NY   Lesson 14   5•1

Concept Development (30 minutes)

Materials: (S) Place value chart, disks for each student

Problem 1

6.72 ÷ 3 = ___

5.16 ÷ 4 = ___

T: (Write 6.72 ÷ 3 = ___ on the board and draw a place value chart with 3 groups at bottom.) Show 6.72 on your place value chart using the number disks. I'll draw on my chart.

S: (Students represent their work with the disks. For the first problem, the students will show their work with the number disks, and the teacher will represent the work in a drawing. In the next problem set, students may draw instead of using the disks.)

T: Let's begin with our largest units. We will share 6 ones equally with 3 groups. How many ones are in each group?

S: 2 ones. (Students move disks to show distribution.)

T: (Draw 2 disks in each group and cross off in the dividend as they are shared.) We gave each group 2 ones. (Record 2 in the ones place in the quotient.) How many ones did we share in all?

S: 6 ones.

T: (Show subtraction in algorithm.) How many ones are left to share?

S: 0 ones.

T: Let's share our tenths. 7 tenths divided by 3. How many tenths can we share with each group?

S: 2 tenths.

T: Share your tenths disks, and I'll show what we did on my mat and in my written work. (Draw to share, cross off in dividend. Record in the algorithm.)

S: (Students move disks.)

T: (Record 2 in tenths place in the quotient.) How many tenths did we share in all?

S: 6 tenths.

T: (Record subtraction.) Let's stop here a moment. Why are we subtracting the 6 tenths?

S: We have to take away the tenths we have already shared. → We distributed the 6 tenths into 3 groups, so we have to subtract it.

T: Since we shared 6 tenths in all, how many tenths are left to share?

NOTE ON MULTIPLE MEANS OF REPRESENTATION:

In order to activate prior knowledge, have students solve one or two whole number division problems using the number disks. Help them record their work, step-by-step, in the standard algorithm. This may help students understand that division of whole numbers and the division of fractions is the same concept and process.

Lesson 14: Divide decimals with a remainder using place value understanding and relate to a written method.
Date: 6/28/13

1.F.19

S: 1 tenth.
T: Can we share 1 tenth with 3 groups?
S: No.
T: What can we do to keep sharing?
S: We can change 1 tenth for 10 hundredths.
T: Make that exchange on your mat. I'll record.
T: How many hundredths do we have now?
S: 12 hundredths.
T: Can we share 12 hundredths with 3 groups? If so, how many hundredths can we share with each group?
S: Yes. We can give 4 hundredths to each group.
T: Share your hundredths and I'll record.
T: (Record 4 hundredths in quotient.) Each group received 4 hundredths. How many hundredths did we share in all?
S: 12 hundredths.
T: (Record subtraction.) Remind me why we subtract these 12 hundredths? How many hundredths are left?
S: We subtract because those 12 hundredths have been shared. → They are divided into the groups now, so we have to subtract 12 hundredths minus 12 hundredths which is equal to 0 hundredths.
T: Look at the 3 groups you made. How many are in each group?
S: 2 and 24 hundredths.
T: Do we have any other units to share?
S: No.
T: How is the division we did with decimal units like whole number division? Turn and talk.
S: It's the same as dividing whole numbers except we are sharing units smaller than ones. → Our quotient has a decimal point because we are sharing fractional units. The decimal shows where the ones place is. → Sometimes we have to change the decimal units just like changing the whole number units in order to continue dividing.
T: (Write 5.16 ÷ 4 = ___ on board.) Let's switch jobs for this problem. I will use disks. You record using the algorithm.

Follow questioning sequence from above as students record steps of algorithm as teacher works the place value disks.

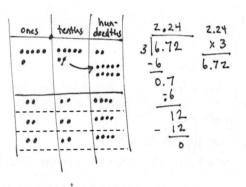

NOTES ON MULTIPLE MEANS OF ACTION AND EXPRESSION:

Students should have the opportunity to use tools that will enhance their understanding. In math class, this often means using manipulatives. Communicate to students that the journey from concrete understanding to representational understanding (drawings) to abstraction is rarely a linear one. Create a learning environment in which students feel comfortable returning to concrete manipulatives when problems are challenging. Throughout this module, the number disks should be readily available to all learners.

Lesson 14: Divide decimals with a remainder using place value understanding and relate to a written method.
Date: 6/28/13

COMMON CORE MATHEMATICS CURRICULUM • NY   **Lesson 14 5•1**

Problem 2

6.72 ÷ 4 = ___

20.08 ÷ 8 = ___

T: (Show 6.72 ÷ 4 = ___ on the board.) Solve this problem using the place value chart with your partner. Partner A will draw the number disks and partner B will record all steps using the standard algorithm.

S: (Students solve.)

T: Compare the drawing to algorithm. Match each number to its counterpart in the drawing.

Circulate to ensure that students are using their whole number experiences with division to share decimal units. Check for misconceptions in recording. For the second problem in the set, partners should switch roles.

Problem 3

6.372 ÷ 6 = ___

T: (Show 6.372 ÷ 6 = ___ on the board.) Work independently using the standard algorithm to solve.

S: (Students solve.)

T: Compare your quotient with your partner. How is this problem different from the ones in the other problem sets? Turn and talk.

Problem Set (10 minutes)

Students should do their personal best to complete the Problem Set within the allotted 10 minutes. For some classes, it may be appropriate to modify the assignment by specifying which problems they work on first. Some problems do not specify a method for solving. Students solve these problems using the RDW approach used for Application Problems.

Student Debrief (10 minutes)

Lesson Objective: Divide decimals with a remainder using place value understanding and relate to a written method.

The Student Debrief is intended to invite reflection and active processing of the total lesson experience.

Invite students to review their solutions for the Problem Set. They should check work by comparing answers

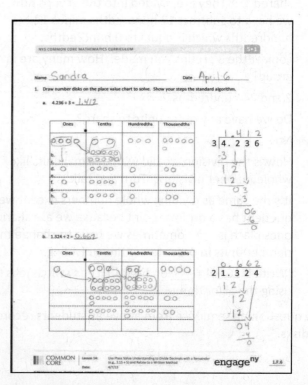

Lesson 14: Divide decimals with a remainder using place value understanding and relate to a written method.
Date: 6/28/13

with a partner before going over answers as a class. Look for misconceptions or misunderstandings that can be addressed in the Debrief. Guide students in a conversation to debrief the Problem Set and process the lesson. You may choose to use any combination of the questions below to lead the discussion.

- How are dividing decimals and dividing whole numbers similar? How are they different?
- Look at the quotients in Problem 1(a) and 1(b). What do you notice about the values in the ones place? Explain why 1b has a zero in the ones place.
- Explain your approach to Problem 4. Because this is a multi-step problem, students may have arrived at the solution through different means. Some may have divided $4.10 by 5 and compared the quotient to the regularly priced avocado. Others may first multiply the regular price, $0.94, by 5, subtract $4.10 from that product, and then divide the difference by 5. Both approaches will result in a correct answer of $0.12 saved per avocado.

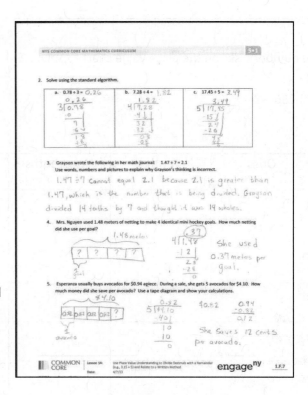

Exit Ticket (3 minutes)

After the Student Debrief, instruct students to complete the Exit Ticket. A review of their work will help you assess the students' understanding of the concepts that were presented in the lesson today and plan more effectively for future lessons. You may read the questions aloud to the students.

COMMON CORE MATHEMATICS CURRICULUM • NY   **Lesson 14 Problem Set**   **5•1**

Name _____   Date _____

1. Draw number disks on the place value chart to solve. Show your steps using the standard algorithm.

    a. 4.236 ÷ 3 = _____

Ones	Tenths	Hundredths	Thousandths

    ```
        _____
     3 | 4 . 2 3 6
    ```

    b.
    c.
    d.
    e.
    f.
    g.
    h.
    i.

    b. 1.324 ÷ 2 = _____

Ones	Tenths	Hundredths	Thousandths

    ```
        _____
     2 | 1 . 3 2 4
    ```

Lesson 14: Divide decimals with a remainder using place value understanding and relate to a written method.

2. Solve using the standard algorithm.

a. $0.78 \div 3 =$ _____	b. $7.28 \div 4 =$ _____	c. $17.45 \div 5 =$ _____

3. Grayson wrote the following in her math journal: $1.47 \div 7 = 2.1$
   Use words, numbers and pictures to explain why Grayson's thinking is incorrect.

4. Mrs. Nguyen used 1.48 meters of netting to make 4 identical mini hockey goals. How much netting did she use per goal?

5. Esperanza usually buys avocados for $0.94 apiece. During a sale, she gets 5 avocados for $4.10. How much money did she save per avocado? Use a tape diagram and show your calculations.

COMMON CORE MATHEMATICS CURRICULUM • NY   Lesson 14 Exit Ticket   5•1

Name _____   Date _____

1. Draw number disks on the place value chart to solve. Show your steps using long division.

   a. 5.372 ÷ 2 = _____

Ones	Tenths	Hundredths	Thousandths

   $2 \overline{)5.372}$

2. Solve using the standard algorithm.

   a. 0.178 ÷ 4 = _____

Lesson 14: Divide decimals with a remainder using place value understanding and relate to a written method.
Date: 6/28/13

COMMON CORE MATHEMATICS CURRICULUM • NY
Lesson 14 Homework 5•1

Name _____ Date _____

1. Draw number disks on the place value chart to solve. Show your steps using long division.

   a. 5.241 ÷ 3 = _____

Ones	Tenths	Hundredths	Thousandths

   $3\overline{)5.241}$

   b. 3.445 ÷ 5 = _____

Ones	Tenths	Hundredths	Thousandths

   $5\overline{)3.445}$

2. Solve using the standard algorithm.

Lesson 14: Divide decimals with a remainder using place value understanding and relate to a written method.
Date: 6/28/13

a. 0.64 ÷ 4 = _____	b. 6.45 ÷ 5 = _____	c. 16.404 ÷ 6 = _____

3. Mrs. Mayuko paid $40.68 for 3 kg of shrimp. What's the cost of 1 kg of shrimp?

4. The total weight of 6 pieces of butter and a bag of sugar is 3.8 lb. If the weight of the bag of sugar is 1.4 lb., what's the weight of each piece of butter?

Lesson 15

Objective: Divide decimals using place value understanding, including remainders in the smallest unit.

Suggested Lesson Structure

- **Fluency Practice** (12 minutes)
- **Application Problems** (8 minutes)
- **Concept Development** (30 minutes)
- **Student Debrief** (10 minutes)
- **Total Time** **(60 minutes)**

Fluency Practice (12 minutes)

- Multiply by Exponents **5.NBT.2**   (8 minutes)
- Find the Quotient **5.NBT.7**   (4 minutes)

Sprint: Multiply by Exponents (8 minutes)

Materials: (S) Multiply by Exponents Sprint

Note: This Sprint will help students build automaticity in multiplying decimals by 10^1, 10^2, 10^3, and 10^4.

Find the Quotient (4 minutes)

Materials: (S) Personal white boards with place value chart

Note: This review fluency will help students work towards mastery of dividing decimal concepts introduced in Lesson 14.

T: (Project place value chart showing ones, tenths, and hundredths. Write 0.48 ÷ 2 = ___.) On your place value chart, draw 48 hundredths in number disks.

S: (Students draw.)

T: (Write 48 hundredths ÷ 2 = ___ hundredths = ___ tenths ___ hundredths.) Solve the division problem.

S: Students write 48 hundredths ÷ 2 = 24 hundredths = 2 tenth 4 hundredths.

T: Now solve using the standard algorithm.

Repeat the process for 0.42 ÷ 3, 3.52 ÷ 2, and 96 tenths ÷ 8.

Lesson 15   5•1

Application Problem (8 minutes)

Jose bought a bag of 6 oranges for $2.82. He also bought 5 pineapples. He gave the cashier $20 and received $1.43 change. What did each pineapple cost?

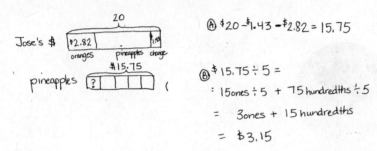

Note: This multi-step problem requires several skills taught in Module 1: multiplying a decimal number by a single-digit, subtraction of decimals numbers, and finally, division of a decimal number. This helps activate prior knowledge that will help scaffold today's lesson on decimal division. Teachers may choose to support students by doing the tape diagram together in order to help students contextualize the details in the story problem.

NOTES ON MULTIPLE MEANS OF REPRESENTATION:

Tape diagrams are a form of modeling that offers students a way to organize, prioritize, and contextualize information in story problems. Students create pictures, represented in bars, from the words in the story problems. Once bars are drawn and the unknown identified, students can find viable solutions.

Concept Development (30 minutes)

Materials: (S) Place value chart

Problems 1–2

$1.7 \div 2$

$2.6 \div 4$

T: (Write $1.7 \div 2$ on the board, and draw a place value chart.) Show 1.7 on your place value chart by drawing number disks. (For this problem, students are only using the place value chart and drawing the number disks. However, the teacher should record the standard algorithm in addition to drawing the number disks, as each unit is decomposed and shared.)

S: (Students draw.)

T: Let's begin with our largest units. Can 1 one be divided into 2 groups?

S: No.

T: Each group gets how many ones?

NOTES ON MULTIPLE MEANS OF REPRESENTATION:

In this lesson students will need to know that a number can be written in multiple ways. In order to activate prior knowledge and heighten interest, the teacher may display a dollar bill, while writing $1 on the board. The class could discuss that in order for the dollar to be divided between two people, it must be thought of as tenths: $1.0. Additionally, if the dollar were to be divided by more than 10 people, it would be thought of as hundredths: $1.00. If students need additional support, this could be demonstrated using concrete materials.

Lesson 15: Divide decimals using place value understanding, including remainders in the smallest unit.
Date: 6/28/13

S: 0 ones.
T: (Record 0 in the ones place of the quotient.) We need to keep sharing. How can we share this single one disk?
S: Unbundle it or exchange it for 10 tenths.
T: Draw that unbundling and tell me how many tenths we have now.
S: 17 tenths.
T: 17 tenths divided by 2. How many tenths can we put in each group?
S: 8 tenths.
T: Cross them off as you divide them into our 2 equal groups.
S: (Students cross out tenths and share them in 2 groups.)
T: (Record 8 tenths in the quotient.) How many tenths did we share in all?
S: 16 tenths.
T: Explain to your partner why we are subtracting the 16 tenths?
S: (Students share.)
T: How many tenths are left?
S: 1 tenth.
T: Is there a way for us to keep sharing? Turn and talk.
S: We can make 10 hundredths with 1 tenth. → Yes, our 1 tenth is still equal to 10 hundredths, even though there is no digit in the hundredths place in 1.7 → We can think about 1 and 7 tenths as 1 and 70 hundredths. It's equal.
T: You unbundle the 1 tenth to make 10 hundredths.
S: (Students unbundle and draw.)
T: Have you changed the value of what we needed to share? Explain.
S: No, it's the same amount to share, but we are using smaller units. → The value is the same - 1 tenth is the same as 10 hundredths.
T: I can show this by placing a zero in the hundredths place.
T: Now that we have 10 hundredths, can we divide this between our 2 groups? How many hundredths are in each group?
S: Yes, 5 hundredths in each group.
T: Let's cross them off as you divide them into 2 equal groups.
S: (Students cross out hundredths and share.)
T: (Record 5 hundredths in the quotient.) How many hundredths did we share in all?
S: 10 hundredths.
T: How many hundredths are left?
S: 0 hundredths.
T: Do we have any other units that we need to share?

Lesson 15: Divide decimals using place value understanding, including remainders in the smallest unit.
Date: 6/28/13

S: No.

T: Tell me the quotient in unit form and in standard form.

S: 0 ones 8 tenths 5 hundredths; 85 hundredths; 0.85

T: (Show 6.72 ÷ 3 = 2.24 in the standard algorithm and 1.7 ÷ 2 = 0.85 in standard algorithm side by side.) How is today's problem different than yesterday's problem? Turn and share with your partner.

S: One problem is divided by 3 and the other one is divided by 2. → Both quotients have 2 decimal places. Yesterday's dividend was to the hundredths, but today's dividend is to the tenths. → We had to think about our dividend as 1 and 70 hundredths to keep sharing. → In yesterday's problem, we had smaller units to unbundle. Today we had smaller units to unbundle, but we couldn't see them in our dividend at first.

T: That's right! In today's problem, we had to record a zero in the hundredths place to show how we unbundled. Did recording that zero change the amount that we had to share – 1 and 7 tenths? Why or why not?

S: No, because 1 and 70 hundredths is the same amount as 1 and 7 tenths.

For the next problem (2.6 ÷ 4) repeat this sequence having students record steps of algorithm as teacher works the mat. Stop along the way to make connections between the concrete materials and the written method.

Problems 3–4

17 ÷ 4

22 ÷ 8

T: (Show 17 ÷ 4 = ____ on the board.) Look at this problem; what do you notice? Turn and share with your partner.

S: When we divide 17 into 4 groups, we will have a remainder.

T: In fourth grade we recorded this remainder as *r1*. What have we done today that lets us keep sharing this remainder?

S: We can trade it for tenths or hundredths and keep dividing it between our groups.

T: Now solve this problem using the place value chart with your partner. Partner A will draw the number disks and Partner B will solve using the standard algorithm.

S: (Students solve.)

T: Compare your work. Match each number in the algorithm with its counterpart in the drawing.

Circulate to ensure that students are using their whole number experiences with division to share decimal units. Check for misconceptions in recording. For the second problem in the set, partners should switch roles.

Lesson 15: Divide decimals using place value understanding, including remainders in the smallest unit.

Date: 6/28/13

COMMON CORE MATHEMATICS CURRICULUM • NY   Lesson 15   5•1

Problem 5

7.7 ÷ 4

T: (Show 7.7 ÷ 4 = on the board.) This time work independently using the standard algorithm to solve.

S: (Students solve.)

T: Compare your answer with your neighbor.

Problem 6

0.84 ÷ 4

T: (Show 0.84 ÷ 4 = on the board.) This time work independently using the standard algorithm to solve.

S: (Students solve.)

T: Compare your answer with your neighbor.

Problem Set (10 minutes)

Students should do their personal best to complete the Problem Set within the allotted 10 minutes. For some classes, it may be appropriate to modify the assignment by specifying which problems they work on first. Some problems do not specify a method for solving. Students solve these problems using the RDW approach used for Application Problems.

Student Debrief (10 minutes)

Lesson Objective: Divide decimals using place value understanding, including remainders in the smallest unit.

The Student Debrief is intended to invite reflection and active processing of the total lesson experience.

Invite students to review their solutions for the Problem Set. They should check work by comparing answers with a partner before going over answers as a class. Look for misconceptions or misunderstandings that can be addressed in the Debrief. Guide students in a conversation to debrief the Problem Set and process the lesson. You may choose to use any combination of the questions below to lead the discussion.

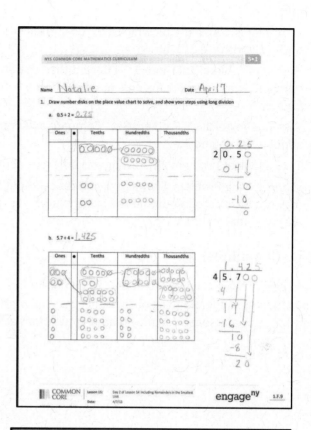

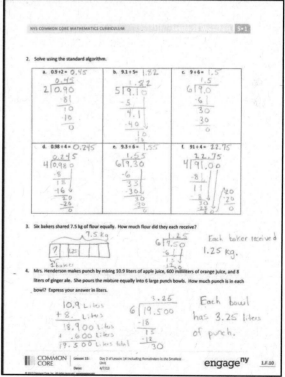

Lesson 15: Divide decimals using place value understanding, including remainders in the smallest unit.
Date: 6/28/13

- In Problems 1(a) and 1(b), which division strategy do you find more efficient? Drawing number disks or the algorithm?
- How are Problems 2(c) and 2(f) different than the others? Will a whole number divided by a whole number always result in a whole number? Explain why these problems resulted in a decimal quotient.
- Take out yesterday's Problem Set. Compare and contrast the first page of each assignment. Talk about what you notice.
- Take a look at Problem 2(f). What was different about how you solved this problem?

When solving Problem 4, what did you notice about the units used to measure the juice? (Students may not have recognized that the orange juice was measured in milliliters.) How do we proceed if we have unlike units?

Exit Ticket (3 minutes)

After the Student Debrief, instruct students to complete the Exit Ticket. A review of their work will help you assess the students' understanding of the concepts that were presented in the lesson today and plan more effectively for future lessons. You may read the questions aloud to the students.

Lesson 15: Divide decimals using place value understanding, including remainders in the smallest unit.
Date: 6/28/13

COMMON CORE MATHEMATICS CURRICULUM • NY   **Lesson 15 Sprint** **5•1**

A   # Correct _____
Solve.

#	Problem		#	Problem	
1	10 × 10 =		23	24 × 10² =	
2	10² =		24	24.7 × 10² =	
3	10² × 10 =		25	24.07 × 10² =	
4	10³ =		26	24.007 × 10² =	
5	10³ × 10 =		27	53 × 1000 =	
6	10⁴ =		28	53 × 10³ =	
7	3 × 100 =		29	53.8 × 10³ =	
8	3 × 10² =		30	53.08 × 10³ =	
9	3.1 × 10² =		31	53.082 × 10³ =	
10	3.15 × 10² =		32	9.1 × 10,000 =	
11	3.157 × 10² =		33	9.1 × 10⁴ =	
12	4 × 1000 =		34	91.4 × 10⁴ =	
13	4 × 10³ =		35	9.104 × 10⁴ =	
14	4.2 × 10³ =		36	9.107 × 10⁴ =	
15	4.28 × 10³ =		37	1.2 × 10² =	
16	4.283 × 10³ =		38	0.35 × 10³ =	
17	5 × 10,000 =		39	5.492 × 10⁴ =	
18	5 × 10⁴ =		40	8.04 × 10³ =	
19	5.7 × 10⁴ =		41	7.109 × 10⁴ =	
20	5.73 × 10⁴ =		42	0.058 × 10² =	
21	5.731 × 10⁴ =		43	20.78 × 10³ =	
22	24 × 100 =		44	420.079 × 10² =	

© Bill Davidson

Lesson 15: Divide decimals using place value understanding, including remainders in the smallest unit.
Date: 6/28/13

B Improvement _____   # Correct _____

Solve.

#	Problem		#	Problem	
1	10 x 10 x 1 =		23	42 x 10^2 =	
2	10^2 =		24	42.7 x 10^2 =	
3	10^2 x 10 =		25	42.07 x 10^2 =	
4	10^3 =		26	42.007 x 10^2 =	
5	10^3 x 10 =		27	35 x 1000 =	
6	10^4 =		28	35 x 10^3 =	
7	4 x 100 =		29	35.8 x 10^3 =	
8	4 x 10^2 =		30	35.08 x 10^3 =	
9	4.1 x 10^2 =		31	35.082 x 10^3 =	
10	4.15 x 10^2 =		32	8.1 x 10,000 =	
11	4.157 x 10^2 =		33	8.1 x 10^4 =	
12	5 x 1000 =		34	81.4 x 10^4 =	
13	5 x 10^3 =		35	8.104 x 10^4 =	
14	5.2 x 10^3 =		36	8.107 x 10^4 =	
15	5.28 x 10^3 =		37	1.3 x 10^2 =	
16	5.283 x 10^3 =		38	0.53 x 10^3 =	
17	7 x 10,000 =		39	4.391 x 10^4 =	
18	7 x 10^4 =		40	7.03 x 10^3 =	
19	7.5 x 10^4 =		41	6.109 x 10^4 =	
20	7.53 x 10^4 =		42	0.085 x 10^2 =	
21	7.531 x 10^4 =		43	30.87 x 10^3 =	
22	42 x 100 =		44	530.097 x 10^2 =	

© Bill Davidson

Lesson 15: Divide decimals using place value understanding, including remainders in the smallest unit.
Date: 6/28/13

COMMON CORE MATHEMATICS CURRICULUM • NY   Lesson 15 Problem Set   5•1

Name _____   Date _____

1. Draw number disks on the place value chart to solve, and show your steps using long division.

   a. 0.5 ÷ 2 = _____

Ones	•	Tenths	Hundredths	Thousandths

   $2\overline{)0.5}$

   b. 5.7 ÷ 4 = _____

Ones	•	Tenths	Hundredths	Thousandths

   $4\overline{)5.7}$

Lesson 15: Divide decimals using place value understanding, including remainders in the smallest unit.
Date: 6/28/13

2. Solve using the standard algorithm.

a. 0.9 ÷ 2 =	b. 9.1 ÷ 5 =	c. 9 ÷ 6 =
d. 0.98 ÷ 4 =	e. 9.3 ÷ 6 =	f. 91 ÷ 4 =

3. Six bakers shared 7.5 kg of flour equally. How much flour did they each receive?

4. Mrs. Henderson makes punch by mixing 10.9 liters of apple juice, 600 milliliters of orange juice, and 8 liters of ginger ale. She pours the mixture equally into 6 large punch bowls. How much punch is in each bowl? Express your answer in liters.

Name _____ Date _____

1. Draw number disks on the place value chart to solve, and show your steps using long division.

   0.9 ÷ 4 = _____

Ones	•	Tenths	Hundredths	Thousandths

   4) 0 . 9

2. Solve using the standard algorithm.

   9.8 ÷ 5 =

COMMON CORE MATHEMATICS CURRICULUM • NY   Lesson 15 Homework   5•1

Name _____   Date _____

1. Draw number disks on the place value chart to solve, and show your steps using long division.

    a.   0.7 ÷ 4 = _____

Ones	•	Tenths	Hundredths	Thousandths

$$4\overline{)0.7}$$

    b.   8.1 ÷ 5 = _____

Ones	•	Tenths	Hundredths	Thousandths

$$5\overline{)8.1}$$

Lesson 15: Divide decimals using place value understanding, including remainders in the smallest unit.
Date: 6/28/13

2. Solve using the standard algorithm.

a. 0.7 ÷ 2 =	b. 3.9 ÷ 6 =	c. 9 ÷ 4 =
d. 0.92 ÷ 2 =	e. 9.4 ÷ 4 =	f. 91 ÷ 8 =

3. A rope 8.7 m long is cut into 5 equal pieces. How long is each piece?

4. Yasmine bought 6 gallons of apple juice. After filling up 4 bottles of the same size with apple juice, she had 0.3 gallon of apple juice left. What's the amount of apple juice in each bottle?

Lesson 16

Objective: Solve word problems using decimal operations.

Suggested Lesson Structure

- ■ Fluency Practice   (12 minutes)
- ■ Application Problems (7 minutes)
- ■ Concept Development   (31 minutes)
- ■ Student Debrief   (10 minutes)
- **Total Time**   **(60 minutes)**

Fluency Practice (12 minutes)

- Divide by Exponents **5.NBT.2**   (8 minutes)
- Find the Quotient   **5.NBT.7**   (4 minutes)

Sprint: Divide by Exponents (8 minutes)

Materials: (S) Divide by Exponents Sprint

Note: This Sprint will help students build automaticity in dividing decimals by 10^1, 10^2, 10^3, and 10^4.

Find the Quotient (4 minutes)

Materials: (S) Personal white boards with place value chart

Note: This review fluency will help students work towards mastery of dividing decimal concepts introduced in Lesson 15.

- T: (Project place value chart showing ones, tenths, and hundredths. Write 0.3 ÷ 2 = ___.) On your place value chart, draw 3 tenths in number disks.
- S: (Students draw.)
- T: (Write 3 tenths ÷ 2 = ___ hundredths ÷ 2 = ___ tenths ___ hundredths on the board.) Solve the division problem.
- S: (Students write 3 tenths ÷ 2 = 30 hundredths ÷ 2 = 1 tenth 5 hundredths.)
- T: (Write the algorithm below 3 tenths ÷ 2 = 30 hundredths ÷ 2 = 1 tenth 5 hundredths.) Solve using the standard algorithm.
- S: (Students solve.)

Repeat process for 0.9 ÷ 5; 6.7 ÷ 5; 0.58 ÷ 4; and 93 tenths ÷ 6.

COMMON CORE MATHEMATICS CURRICULUM • NY   Lesson 16   5•1

Application Problems (7 minutes)

Jesse and three friends buy snacks for a hike. They buy trail mix for $5.42, apples for $2.55, and granola bars for $3.39. If the four friends split the cost of the snacks equally, how much should each friend pay?

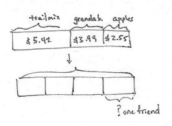

Note: Adding and dividing decimals are taught in this module. Teachers may choose to help students draw the tape diagram before students do the calculations independently.

Concept Development (31 minutes)

Materials: (T/S) Problem Set, pencils

Problem 1

Mr. Frye distributed $126 equally among his 4 children for their weekly allowance. How much money did each child receive?

As the teacher creates each component of the tape diagram, students should recreate the tape diagram on their problem set.

- T: We will work Problem 1 on your Problem Set together. (Project problem on board.) Read the word problem together.
- S: (Students read chorally.)
- T: Who and what is this problem about? Let's identify our variables.
- S: Mr. Frye's money.
- T: Draw a bar to represent Mr. Frye's money.

Mr. Frye's money   [   ]

- T: Let's read the problem sentence by sentence and adjust our diagram to match the information in the problem. Read the first sentence together.
- S: (Students read.)
- T: What is the important information in the first sentence? Turn and talk.

Lesson 16:   Solve word problems using decimal operations.
Date: 6/28/13

1.F.42

S: $126 and 4 children received an equal amount.
T: (Underline stated information.) How can I represent this information in my diagram?
S: 126 dollars is the total, so put a bracket on top of the bar and label it.
T: (Draw a bracket over the diagram and label as $126. Have students label their diagram.)

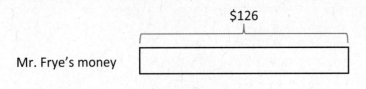

T: How many children share the 126 dollars?
S: 4 children.
T: How can we represent this information?
S: Divide the bar into 4 equal parts.
T: (Partition the diagram into 4 sections and have students do the same.)

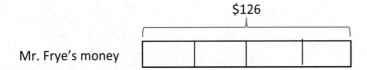

NOTES ON MULTIPLE MEANS OF REPRESENTATION:

Students may use various approaches for calculating the quotient. Some may use place value units 12 tens + 60 tenths. Others may use the division algorithm. Discussion focusing on comparisons between and among approaches to computation supports students in becoming strategic mathematical thinkers.

T: What is the question?
S: How much did each child receive?
T: What is unknown in this problem? How will we represent it in our diagram?
S: The amount of money one of Mr. Frye's children received for allowance is what we are trying to find. We should put a question mark inside one of the parts.
T: (Write a question mark inside of each part of the tape diagram.)

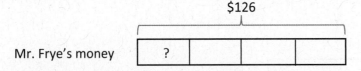

NOTES ON MULTIPLE MEANS OF ENGAGEMENT:

If students struggle to draw a model of word problems involving division with decimal values, scaffold their understanding by modeling an analogous problem substituting simpler, whole number values. Then using the same tape diagram, erase the whole number values and replace them with the parallel value from the decimal problem.

T: Make a unit statement about your diagram. (Alternately – How many unit bars are equal to $126?)
S: 4 units is the same as $126.
T: How can we find the value of one unit?
S: Divide $126 by 4. → Use division, because we have a whole that we are sharing equally.
T: What is the equation that will give us the amount that each child receives?

Lesson 16: Solve word problems using decimal operations.
Date: 6/28/13

COMMON CORE MATHEMATICS CURRICULUM • NY   Lesson 16   5•1

S: $126 ÷ 4 = _____.
T: Solve and express your answer in a complete sentence.
S: (Students work.)

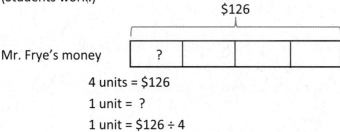

Mr. Frye's money

4 units = $126
1 unit = ?
1 unit = $126 ÷ 4
       = $31.50

S: Each child received $31.50 for their weekly allowance.
T: Look at part b of question 1 and solve using a tape diagram.
S: (Students work for 5 minutes.)

As students are working, circulate and be attentive to accuracy and labeling of information in the tape diagram. Also see student sample of the Problem Set for possible diagrams.

Problem 2

Brandon mixed 6.83 lbs. of cashews with 3.57 lbs. of pistachios. After filling up 6 bags that were the same size with the mixture, he had 0.35 lbs. of nuts left. What was the weight of each bag?

T: Read the problem. Identify the variables (who and what) and draw a bar.
S: (Students read and draw.)

Brandon's cashews/pistachios

T: Read the first sentence.
S: (Students read.)
T: What is the important information in this sentence? Tell a partner.
S: 6.83 lbs. of cashews and 3.57 lbs. of pistachios.
T: (Underline the stated information.) How can I represent this information in our tape diagram?
S: Show two parts inside the bar.
T: Should the parts be equal in size?
S: No. The cashews part should be about twice the size of the pistachio part.

MP.8

         6.83   3.87

Brandon's cashews/pistachios

Lesson 16:   Solve word problems using decimal operations.
Date: 6/28/13

T: (Draw and label.) Let's read the next sentence. How will we represent this part of the problem?

S: We could draw another bar to represent both kinds of nuts together and split it into parts to show the bags and the part that was left over. → We could erase the bar separating the nuts, put the total on the bar we already drew and split it into the equal parts, but we have to remember he had some nuts left over.

T: Both are good ideas, choose one for your model. I am going to use the bar that I've already drawn. I'll label my bags with the letter b and label the part that wasn't put into a bag.

T: (Erase the bar between the types of nuts. Draw a bracket over the bar and write the total. Show the left over nuts and the 6 bags.)

Brandon's cashews/pistachios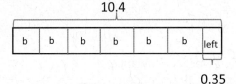

T: What is the question?
S: How much did each bag weigh?
T: Where should we put our question mark?
S: Inside one of the units that is labeled with the letter b.

Brandon's cashews/pistachios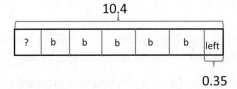

T: How will we find the value of 1 unit in our diagram? Turn and talk.

S: Part of the weight is being placed into 6 bags, we need to divide that part by 6. → There was a part that didn't get put in a bag. We have to take the left over part away from the total so we can find the part that was divided into the bags. Then we can divide.

T: Perform your calculations and state your answer in a complete sentence. (Please see above for solution.)

NOTES ON MULTIPLE MEANS OF REPRESENTATION:

Complex relationships within a tape diagram can be made clearer to students with the use of color. The bags of cashews in Problem 4 could be made more visible by outlining the bagged nuts in red. This creates a classic part-part-whole problem. Students can readily see the portion that must be subtracted in order to produce the portion divided into 6 bags.

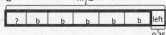

If using color to highlight relationships is still too abstract for students, colored paper can be cut, marked, and manipulated.

"Thinking Blocks" is a free internet site which offers students with fine motor deficits a tool for drawing bars and labels electronically. Models can be printed for sharing with classmates.

Lesson 16: Solve word problems using decimal operations.
Date: 6/28/13

COMMON CORE MATHEMATICS CURRICULUM • NY   Lesson 16 5•1

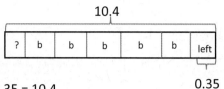

Brandon's cashews/pistachios

6 units + 0.35 = 10.4

1 unit = (10.4 − 0.35) ÷ 6

1 unit = 1.675 lbs

Each bag contained 1.675 lbs of nuts.

T: Complete questions 2, 3, and 5 on the worksheet, using a tape diagram and calculations to solve.

Circulate as students work, listening for sound mathematical reasoning.

Problem Set (please see note below)

Today's problem set forms the basis of the Concept Development. Students will work Problems 1 and 4 with teacher guidance, modeling and scaffolding. Problems 2, 3, and 5 are designed to be independent work for the last 15 minutes of concept development.

NOTES ON MULTIPLE MEANS OF REPRESENTATION:

The equations pictured to the right are a formal teacher solution for Question 4. Students should not be expected to produce such a formal representation of their thinking. Students are more likely to simply show a vertical subtraction of the left over nuts from the total and then show a division of the bagged nuts into 6 equal portions. There may be other appropriate strategies for solving offered by students as well.

Teacher solutions offer an opportunity to expose students to more formal representations. These solutions might be written on the board as a way to translate a student's approach to solving as the student communicates their strategy aloud to the class.

Student Debrief (10 minutes)

Lesson Objective: Solve word problems using decimal operations.

The Student Debrief is intended to invite reflection and active processing of the total lesson experience.

Invite students to review their solutions for the Problem Set. They should check work by comparing answers with a partner before going over answers as a class. Look for misconceptions or misunderstandings that can be addressed in the Debrief. Guide students in a conversation to debrief the problem set and process the lesson. You may choose to use any combination of the questions below to lead the discussion.

- In Question 3, how did you represent the information using the tape diagram?
- How did the tape diagram in 1(a) help you solve 1(b)?

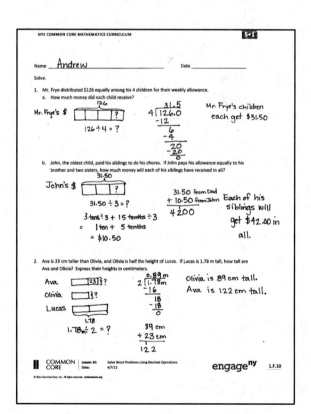

Lesson 16: Solve word problems using decimal operations.
Date: 6/28/13

COMMON CORE MATHEMATICS CURRICULUM • NY   Lesson 16 5•1

- Look at 1(b) and 5(b). How are the questions different? (1(b) is partitive division—groups are known, size of group is unknown. 5(b) is measurement division – size of group is known, number of groups is unknown.) Does the difference in the questions affect the calculation of the answer?

- As an extension or an option for early finishers, have students generate word problems based on labeled tape diagrams and/or have them create one of each type of division problem (group known and group unknown).

Exit Ticket (3 minutes)

After the Student Debrief, instruct students to complete the Exit Ticket. A review of their work will help you assess the students' understanding of the concepts that were presented in the lesson today and plan more effectively for future lessons. You may read the questions aloud to the students.

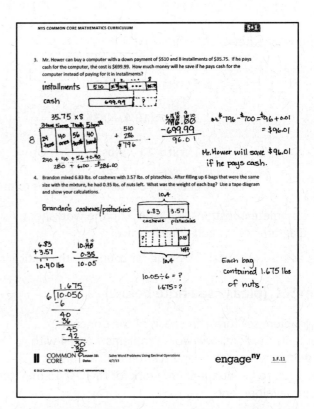

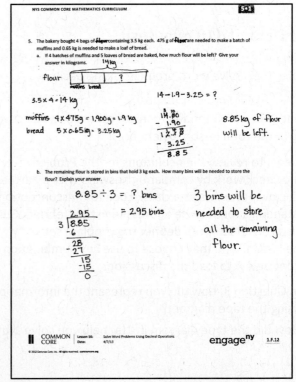

Lesson 16: Solve word problems using decimal operations.
Date: 6/28/13

A # Correct _____

Solve.

#	Problem		#	Problem	
1	10 × 10 =		23	3,400 ÷ 10^2 =	
2	10^2 =		24	3,470 ÷ 10^2 =	
3	10^2 × 10 =		25	3,407 ÷ 10^2 =	
4	10^3 =		26	3,400.7 ÷ 10^2 =	
5	10^3 × 10 =		27	63,000 ÷ 1000 =	
6	10^4 =		28	63,000 ÷ 10^3 =	
7	3 × 100 =		29	63,800 ÷ 10^3 =	
8	3 × 10^2 =		30	63,080 ÷ 10^3 =	
9	3.1 × 10^2 =		31	63,082 ÷ 10^3 =	
10	3.15 × 10^2 =		32	81,000 ÷ 10,000 =	
11	3.157 × 10^2 =		33	81,000 ÷ 10^4 =	
12	4 × 1000 =		34	81,400 ÷ 10^4 =	
13	4 × 10^3 =		35	81,040 ÷ 10^4 =	
14	4.2 × 10^3 =		36	91,070 ÷ 10^4 =	
15	4.28 × 10^3 =		37	120 ÷ 10^2 =	
16	4.283 × 10^3 =		38	350 ÷ 10^3 =	
17	5 × 10,000 =		39	45,920 ÷ 10^4 =	
18	5 × 10^4 =		40	6,040 ÷ 10^3 =	
19	5.7 × 10^4 =		41	61,080 ÷ 10^4 =	
20	5.73 × 10^4 =		42	7.8 ÷ 10^2 =	
21	5.731 × 10^4 =		43	40,870 ÷ 10^3 =	
22	24 × 100 =		44	52,070.9 ÷ 10^2 =	

© Bill Davidson

Lesson 16: Solve word problems using decimal operations.
Date: 6/28/13

B Improvement _____ # Correct _____

Solve.

#	Problem		#	Problem	
1	10 x 10 x 1 =		23	4,370 ÷ 10^2 =	
2	10^2 =		24	4,370 ÷ 10^2 =	
3	10^2 x 10 =		25	4,307 ÷ 10^2 =	
4	10^3 =		26	4,300.7 ÷ 10^2 =	
5	10^3 x 10 =		27	73,000 ÷ 1000 =	
6	10^4 =		28	73,000 ÷ 10^3 =	
7	500 ÷ 100 =		29	73,800 ÷ 10^3 =	
8	500 ÷ 10^2 =		30	73,080 ÷ 10^3 =	
9	510 ÷ 10^2 =		31	73,082 ÷ 10^3 =	
10	516 ÷ 10^2 =		32	91,000 ÷ 10,000 =	
11	516.7 ÷ 10^2 =		33	91,000 ÷ 10^4 =	
12	6,000 ÷ 1000 =		34	91,400 ÷ 10^4 =	
13	6,000 ÷ 10^3 =		35	91,040 ÷ 10^4 =	
14	6,200 ÷ 10^3 =		36	81,070 ÷ 10^4 =	
15	6,280 ÷ 10^3 =		37	170 ÷ 10^2 =	
16	6,283 ÷ 10^3 =		38	450 ÷ 10^3 =	
17	70,000 ÷ 10,000 =		39	54,920 ÷ 10^4 =	
18	70,000 ÷ 10^4 =		40	4,060 ÷ 10^3 =	
19	76,000 ÷ 10^4 =		41	71,080 ÷ 10^4 =	
20	76,300 ÷ 10^4 =		42	8.7 ÷ 10^2 =	
21	76,310 ÷ 10^4 =		43	60,470 ÷ 10^3 =	
22	4,300 ÷ 100 =		44	72,050.9 ÷ 10^2 =	

© Bill Davidson

Lesson 16: Solve word problems using decimal operations.
Date: 6/28/13

Name _____ Date _____

Solve.

1. Mr. Frye distributed $126 equally among his 4 children for their weekly allowance.
   a. How much money did each child receive?

   b. John, the oldest child, paid his siblings to do his chores. If John pays his allowance equally to his brother and two sisters, how much money will each of his siblings have received in all?

2. Ava is 23 cm taller than Olivia, and Olivia is half the height of Lucas. If Lucas is 1.78 m tall, how tall are Ava and Olivia? Express their heights in centimeters.

3. Mr. Hower can buy a computer with a down payment of $510 and 8 monthly payments of $35.75. If he pays cash for the computer, the cost is $699.99. How much money will he save if he pays cash for the computer instead of paying for it in monthly payments?

4. Brandon mixed 6.83 lbs. of cashews with 3.57 lbs. of pistachios. After filling up 6 bags that were the same size with the mixture, he had 0.35 lbs. of nuts left. What was the weight of each bag? Use a tape diagram and show your calculations.

5. The bakery bought 4 bags of flour containing 3.5 kg each. 475 g of flour are needed to make a batch of muffins and 0.65 kg is needed to make a loaf of bread.

   a. If 4 batches of muffins and 5 loaves of bread are baked, how much flour will be left? Give your answer in kilograms.

   b. The remaining flour is stored in bins that hold 3 kg each. How many bins will be needed to store the flour? Explain your answer.

COMMON CORE MATHEMATICS CURRICULUM • NY   Lesson 16 Exit Ticket   5•1

Name _____   Date _____

Write a word problem with two questions that matches the tape diagram below, then solve.

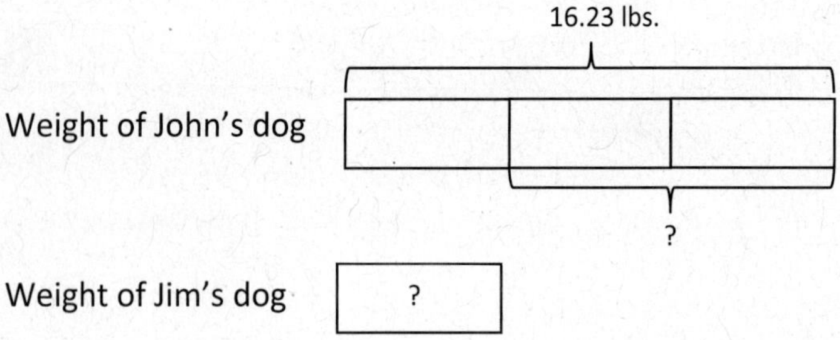

                             16.23 lbs.

Weight of John's dog   [| |]
                          ?

Weight of Jim's dog   [?]

Lesson 16: Solve word problems using decimal operations.
Date: 6/28/13

1.F.53

COMMON CORE MATHEMATICS CURRICULUM • NY

Lesson 16 Homework 5•1

Name _____ Date _____

Solve using tape diagrams.

1. A gardener installed 42.6 meters of fencing in a week. He installed 13.45 meters on Monday and 9.5 meters on Tuesday. He installed the rest of the fence in equal lengths on Wednesday through Friday. How many meters of fencing did he install on each of the last three days?

2. Jenny charges $9.15 an hour to babysit toddlers and $7.45 an hour to babysit school-aged children.

   a. If Jenny babysat toddlers for 9 hours and school-aged children for 6 hours, how much money did she earn in all?

   b. Jenny wants to earn $1300 by the end of the summer. How much more will she need to earn to meet her goal?

Lesson 16: Solve word problems using decimal operations.
Date: 6/28/13

3. A table and 8 chairs weigh 235.68 pounds together. If the table weighs 157.84 lbs., what is the weight of one chair in pounds?

4. Mrs. Cleaver mixes 1.24 liters of red paint with 3 times as much blue paint to make purple paint. She pours the paint equally into 5 containers. How much blue paint is in each cup? Give you answer in liters.

COMMON CORE MATHEMATICS CURRICULUM • NY   Mid-Module Assessment Task   5•1

Name _____   Date _____

1. Compare using >, <, or =.

   a.   0.4   0.127

   b.   2 thousandths + 4 hundredths   0.036

   c.   2 tens 3 tenths 1 thousandth   20.31

   d.   24 tenths   2.5

   e. $4 \times 10^3 + 2 \times 100 + 3 \times \frac{1}{10}$ $4 \times 1000 + 2 \times 10^2 + 3 \times \frac{1}{10}$

   f. $3 \times \frac{1}{10} + 4 \times \frac{1}{1000}$ ◯   0.340

2. Model the number 8.88 on the place value chart.

   a. Use words, numbers, and your model to explain why each of the digits has a different value. Be sure to use "ten times as much" and "one tenth of" in your explanation.

b. Multiply 8.88 x 10⁴. Explain the shift of the digits, the change in the value of each digit, and the number of zeroes in the product.

c. Divide the product from (b) by 10². Explain the shift of the digits and how the value of each digit changed.

3. Rainfall collected in a rain gauge was found to be 2.3 cm when rounded to the nearest tenth of a centimeter.

   a. Circle all the measurements below that could be the actual measurement of the rainfall.

   2.251 cm   2.349 cm   2.352 cm   2.295 cm

   b. Convert the rounded measurement to meters. Write an equation to show your work.

4. Annual rainfall total for cities in New York are listed below.

Rochester	0.97 meters
Ithaca	0.947 meters
Saratoga Springs	1.5 meters
New York City	1.268 meters

   a. Put the rainfall measurements in order from least to greatest. Write the smallest total rainfall in word form and expanded form.

   b. Round each of the rainfall totals to the nearest tenth.

   c. Imagine New York City's rainfall is the same every year. How much rain would fall in 100 years?

   d. Write an equation using an exponent that would express the 100-year total rainfall. Explain how the digits have shifted position and why.

COMMON CORE MATHEMATICS CURRICULUM • NY   Mid-Module Assessment Task 5•1

Mid-Module Assessment Task
Standards Addressed
Topics A–C

Generalize place value understanding for multi-digit whole numbers

5.NBT.1 Recognize that in a multi-digit number, a digit in one place represents 10 times as much as it represents in the place to its right and 1/10 of what it represents in the place to its left.

5.NBT.2. Explain patterns in the number of zeros of the product when multiplying a number by powers of 10, and explain patterns in the placement of the decimal point when a decimal is multiplied or divided by a power of 10. Use whole-number exponents to denote powers of 10.

5.NBT.3 Read, write, and compare decimals to thousandths.

   a. Read and write decimals to thousandths using base-ten numerals, number names, and expanded form, e.g., 347.392 = 3 × 100 + 4 × 10 + 7 × 1 + 3 × (1/10) + 9 × (1/100) + 2 × (1/1000).

   b. Compare two decimals to thousandths based on meanings of the digits in each place, using >, =, and < symbols to record the results of comparisons.

5.NBT.4 Use place value understanding to round decimals to any place.

5.MD.1 Convert among different-sized standard measurement units within a given measurement system (e.g., convert 5 cm to 0.05 m), and use these conversions in solving multi-step, real world problems.

Evaluating Student Learning Outcomes

A Progression Toward Mastery is provided to describe steps that illuminate the gradually increasing understandings that students develop *on their way to proficiency*. In this chart, this progress is presented from left (Step 1) to right (Step 4). The learning goal for each student is to achieve Step 4 mastery. These steps are meant to help teachers and students identify and celebrate what the student CAN do now, and what they need to work on next.

| Module 1: | Place Value and Decimal Fractions |
| Date: | 6/28/13 |

COMMON CORE MATHEMATICS CURRICULUM • NY   Mid-Module Assessment Task   5•1

A Progression Toward Mastery				
Assessment Task Item and Standards Assessed	STEP 1 Little evidence of reasoning without a correct answer. (1 Point)	STEP 2 Evidence of some reasoning without a correct answer. (2 Points)	STEP 3 Evidence of some reasoning with a correct answer or evidence of solid reasoning with an incorrect answer. (3 Points)	STEP 4 Evidence of solid reasoning with a correct answer. (4 Points)
1 5.NBT.3a 5.NBT.3b	The student answers none or 1 part correctly.	The student answers 2 or 3 parts correctly.	The student answers 4 or 5 parts correctly.	The student correctly answers all 6 parts: a. > d. < b. > e. = c. < f. <
2 5.NBT.1 5.NBT.2	The student answers none or 1 part correctly.	The student answers 2 parts correctly.	The student is able to answers all parts correctly but is unable to explain his strategy in (a), (b), or (c), or answers 3 of the 4 parts correctly.	The student accurately models 8.88 on the place value chart, and correctly: • Uses words, numbers, and model to explain why each digit has a different value. • Finds product 88,800 and explains. • Finds quotient of 888 and explains.
3 5.NBT.4 5.MD.1	The student is unable to identify any answers for (a), or answer (b) correctly.	The student identifies 1 or 2 answers correctly for (a), and makes an attempt to convert but gets an incorrect solution for (b).	The student identifies 2 answers correctly for (a), and converts correctly for (b), or identifies 3 answers correctly for (a) and converts with a small error for (b).	The student identifies all 3 answers correctly for (a), and answers (b) correctly: a. 2.251 cm, 2.349 cm, 2.3955 cm. b. $2.3 \times 10^2 = 0.023$ m.

Module 1:   Place Value and Decimal Fractions
Date: 6/28/13

A Progression Toward Mastery				
4 **5.NBT.1** **5.NBT.2** **5.NBT.3** **5.NBT.4**	The student answers none or 1 part correctly.	The student answers 2 problems correctly.	The student is able to answer all parts correctly but is unable to explain strategy in (d), or answers 3 of the 4 problems correctly.	The student correctly responds: a. 0.947 m, 0.97 m, 1.268 m, 1.5 m. • 947 thousandths meters. • 0.9 + 0.04 + 0.007 = 0.947 m. b. Rochester ≈ 1.0 m, Ithaca ≈ 0.9 m, Saratoga Springs ≈ 1.5 m, NYC ≈ 1.3 m. c. 126.8 m. d. 1.268 × 10^2 = 126.8.

COMMON CORE MATHEMATICS CURRICULUM • NY   Mid-Module Assessment Task   **5•1**

Name __Kate_____   Date _____

1. Compare using >, <, or =.

   a. 0.4   0.127

   b. 2 thousandths + 4 hundredths (>) 0.036

   c. 2 tens 3 tenths 1 thousandth (<) 20.31

   d. 24 tenths (<) 2.5

   e. $4 \times 10^3 + 2 \times 100 + 3 \times \frac{1}{10}$ $4 \times 1000 + 2 \times 10^2 + 3 \times \frac{1}{10}$

   f. $3 \times \frac{1}{10} + 4 \times \frac{1}{1000}$ (<) 0.340

2. Model the number 8.88 on the place value chart.

   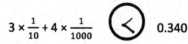

   a. Use words, numbers, and your model to explain why each of the digits has a different value. Be sure to use "ten times as much" and "one tenth of" in your explanation.

   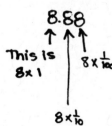

   Even though there are 8 disks in each column, they are different units so they have different values.
   8 ones is 10 times as much as 8 tenths.
   8 hundredths is one tenth as much as 8 tenths.

b. Multiply 8.88 × 10⁴. Explain the shift of the digits, the change in the value of each digit, and the number of zeroes in the product.

$$8 \times \overbrace{10 \times 10 \times 10 \times 10}^{10^4} = 80,000.00$$
$$0.8 \times 10 \times 10 \times 10 \times 10 = 8,000.00$$
$$0.08 \times 10 \times 10 \times 10 \times 10 = 800.00$$
$$88,800.00 \quad \text{you can see 4 zeros.}$$

The digits shift to the left 4 times. It goes one place everytime you multiply by 10. It gets 10× bigger so it has to move to a larger place value.

c. Divide the product from (b) by 10². Explain the shift of the digits and how the value of each digit changed.

$$88,800 \div 10^2 = 888$$

$$80,000 \div 10 \div 10 = 800$$
$$8,000 \div 10 \div 10 = 80$$
$$800 \div 10 \div 10 = 8$$
$$\underline{}$$
$$888$$

3. Rainfall collected in a rain gauge was found to be 2.3 cm when rounded to the nearest tenth of a centimeter.

a. Circle all the measurements below that could be the actual measurement of the rainfall.

(2.251 cm)   (2.349 cm)   2.352 cm   (2.295 cm)

b. Convert the rounded measurement to meters. Write an equation to show your work.

$$2.3 \text{ cm} \div 10^2 = 0.023 \text{ m}$$

4. Annual rainfall total for cities in New York are listed below.

Rochester	0.97 meters
Ithaca	0.947 meters
Saratoga Springs	1.5 meters
New York City	1.268 meters

a. Put the rainfall measurements in order from least to greatest. Write the smallest total rainfall in word form and expanded form.

0.947m, 0.97m, 1.268m, 1.5m

nine hundred forty-seven thousandths

$9 \times \frac{1}{10} + 4 \times \frac{1}{100} + 7 \times \frac{1}{1000}$ $9 \times 0.1 + 4 \times 0.01 + 7 \times 0.001$

b. Round each of the rainfall totals to the nearest tenth.

0.97 ≈ 1.0 m
0.947 ≈ 0.9 m
1.5   1.5 m
1.268 ≈ 1.3 m

c. Imagine New York City's rainfall is the same every year. How much rain would fall in 100 years?

1.268 × 100 = 126.8 m would fall in 100 years.

d. Write an equation using an exponent that would express the 100-year total rainfall. Explain how the digits have shifted position and why.

$1.268 \times 10^2 = 126.8$ m

Each part of the number got 100 times as large so each one moved 2 places to the left. We multiplied by 10 twice so we moved 2 places

1 × 100 = 100
0.2 × 100 = 20
0.06 × 100 = 6
0.008 × 100 = 0.8

Name _____   Date _____

1. The following equations involve different quantities and use different operations, yet produce the same result. Use a place value mat and words to explain why this is true.

    $4.13 \times 10^3 = 4130$ $413{,}000 \div 10^2 = 4130$

2. Use an area model to explain the product of 4.6 and 3. Write the product in standard form, word form and expanded form.

3. Compare using >, <, or =.

   a. 2 tenths + 11 hundredths ◯ 0.13

   b. 13 tenths + 8 tenths + 32 hundredths ◯ 2.42

   c. 342 hundredths + 7 tenths ◯ 3 + 49 hundredths

   d. $2 + 31 \times \frac{1}{10} + 14 \times \frac{1}{100}$ ◯ 2.324

   e. $14 + 72 \times \frac{1}{10} + 4 \times \frac{1}{1000}$ ◯ 21.24

   f. $0.3 \times 10^2 + 0.007 \times 10^3$ ◯ $0.3 \times 10 + 0.7 \times 10^2$

4. Dr. Mann mixed 10.357 g of chemical A, 12.062 g of chemical B, and 7.506 g of chemical C to make 5 doses of medicine.

   a. About how much medicine did he make in grams? Estimate the amount of each chemical by rounding to the nearest tenth of a gram before finding the sum. Show all your thinking.

   b. Find the actual amount of medicine mixed by Dr. Mann. What is the difference in your estimate and the actual amount?

   c. How many grams are in one dose of medicine? Explain your strategy for solving this problem.

   d. Round the weight of one dose to the nearest gram. Write an equation that shows how to convert the rounded weight to kilograms and solve. Explain your thinking in words.

COMMON CORE MATHEMATICS CURRICULUM • NY   End-of-Module Assessment Task   5•1

End-of-Module Assessment Task
Standards Addressed   Topics A–F

Generalize place value understanding for multi-digit whole numbers.

5.NBT.1 Recognize that in a multi-digit number, a digit in one place represents 10 times as much as it represents in the place to its right and 1/10 of what it represents in the place to its left.

5.NBT.2. Explain patterns in the number of zeros of the product when multiplying a number by powers of 10, and explain patterns in the placement of the decimal point when a decimal is multiplied or divided by a power of 10. Use whole-number exponents to denote powers of 10.

5.NBT.3 Read, write, and compare decimals to thousandths.

  a. Read and write decimals to thousandths using base-ten numerals, number names, and expanded form, e.g., 347.392 = 3 × 100 + 4 × 10 + 7 × 1 + 3 × (1/10) + 9 × (1/100) + 2 × (1/1000).

  b. Compare two decimals to thousandths based on meanings of the digits in each place, using >, =, and < symbols to record the results of comparisons.

5.NBT.4 Use place value understanding to round decimals to any place.

Perform operations with multi-digit whole numbers and with decimals to hundredths.

5.NBT.7 Add, subtract, multiply and divide decimals to hundredths, using concrete models or drawings and strategies based on place value, properties of operations, and/or the relationship between addition and subtraction; relate the strategy to a written method and explain the reasoning used.

Convert like measurement units within a given measurement system.

5.MD.1 Convert among different-sized standard measurement units within a given measurement system (e.g., convert 5 cm to 0.05 m), and use these conversions in solving multi-step, real world problems.

Evaluating Student Learning Outcomes

A Progression Toward Mastery is provided to describe steps that illuminate the gradually increasing understandings that students develop *on their way to proficiency.* In this chart, this progress is presented from left (Step 1) to right (Step 4). The learning goal for each student is to achieve Step 4 mastery. These steps are meant to help teachers and students identify and celebrate what the student CAN do now, and what they need to work on next.

Module 1:   Place Value and Decimal Fractions
Date: 6/28/13

End-of-Module Assessment Task 5•1

A Progression Toward Mastery

Assessment Task Item and Standards Assessed	STEP 1 Little evidence of reasoning without a correct answer. (1 Point)	STEP 2 Evidence of some reasoning without a correct answer. (2 Points)	STEP 3 Evidence of some reasoning with a correct answer or evidence of solid reasoning with an incorrect answer. (3 Points)	STEP 4 Evidence of solid reasoning with a correct answer. (4 Points)
1 **5.NBT.1** **5.NBT.2**	The student is unable to provide a correct response.	The student attempts but is not able to accurately draw the place value mat or explain reasoning fully.	The student correctly draws place mat but does not show full reasoning, or explains reasoning fully but place value mat doesn't match the reasoning.	The student correctly: • Draws place value mat showing movement of digits. • Explains movement of units to the left for multiplication and movement of units to the right for division.
2 **5.NBT.7**	The student is unable to use the area model to find the product.	The student attempts using an area model to multiply but inaccurately. Student attempts to write either word or expanded form of inaccurate product	The student uses the area model to multiply but does not find the correct product. Student accurately produces word and expanded form of inaccurate product.	The student correctly: • Draws an area model. • Shows work to find product 13.8. • Accurately expresses product in both word and expanded form.
3 **5.NBT.3a** **5.NBT.3b**	The student answers none or 1 part correctly.	The student answers 2 or 3 answers correctly.	The student answers 4 or 5 answers correctly.	The student correctly answers all 6 parts: a. > d. > b. = e. < c. > f. <

Module 1: Place Value and Decimal Fractions
Date: 6/28/13

A Progression Toward Mastery

4 5.NBT.1 5.NBT.2 5.NBT.3a 5.NBT.3b 5.NBT.4 5.NBT.7 5.MD.1	The student answers none or 1 part correctly.	The student answers 2 problems correctly.	The student is able to find all answers correctly but is unable to explain strategy in (c), or answers 3 of the 4 problems correctly.	The student correctly: a. Estimates 10.357 g to 10.4 g; 12.062g to 12.1 g; and 7.506 as 7.5; finds sum 30 g; shows work or model. b. Finds sum 29.925 g and difference 0.075 g. c. Finds quotient 5.985g and explains accurately strategy used. d. Rounds 5.985g to 6g finds quotient 0.006 kg. Shows equation as 6 ÷ 10^3 or 6 ÷1000 = 0.006kg. Writes either 4 g ÷ 1000 = 0.004kg or 4 g ÷ 10^3 = 0.004 kg.

Module 1: Place Value and Decimal Fractions
Date: 6/28/13

Name __Jackie_____ Date _____

1. The following equations involve different quantities and use different operations, yet produce the same result. Use a place value mat and words to explain why this is true.

   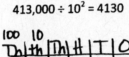

   $4.13 \times 10^3 = 4130$ $413{,}000 \div 10^2 = 4130$

      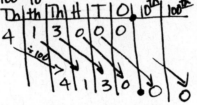

   When I multiplied, the digits moved 3 places to the left because they got larger. When I divided, it moved the digits 2 places to the right because they got smaller.

2. Use an area model to explain the product of 4.6 and 3. Write the product in standard form, word form and expanded form.

   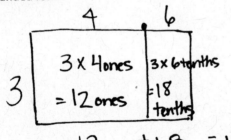

   $12 + 1.8 = 13.8$

   thirteen and 8 tenths

   $1 \times 10 + 3 \times 1 + 8 \times \frac{1}{10}$

3. Compare using >, <, or =.

   a. 2 tenths + 11 hundredths 0.13

   b. 13 tenths + 8 tenths + 32 hundredths 2.42

   c. 342 hundredths + 7 tenths 3 + 49 hundredths

   d. $2 + 31 \times \frac{1}{10} + 14 \times \frac{1}{100}$ 2.324

   e. $14 + 72 \times \frac{1}{10} + 4 \times \frac{1}{1000}$ 21.24

   f. $0.3 \times 10^2 + 0.007 \times 10^3$ $0.3 \times 10 + 0.7 \times 10^2$

4. Dr. Mann mixed 10.357 g of chemical A, 12.062 g of chemical B, and 7.506 g of chemical C to make 5 doses of medicine.

   a. About how much medicine did he make in grams? Estimate the amount of each chemical by rounding to the nearest tenth of a gram before finding the sum. Show all your thinking.

   A 10.357g ≈ 10.4g
   B 12.062g ≈ 12.1g
   C 7.506g ≈ 7.5g

   ```
     10.4
     12.1
   +  7.5
   ──────
     30.0
   ```

   Dr. Mann made about 30g of medicine.

   b. Find the actual amount of medicine mixed by Dr. Mann. What is the difference in your estimate and the actual amount?

   ```
     10.357
     12.062
   +  7.506
   ────────
     29.925
   ```

   ```
     30.000
   - 29.925
   ────────
      0.075
   ```

   The difference in the estimated + actual amounts is 0.075g.

   c. How many grams are in one dose of medicine? Explain your strategy for solving this problem.

   ```
        5.985
      ┌──────
    5 │29.925
      -25
      ───
       49
      -45
      ───
        42
       -40
       ───
        25
       -25
   ```

   I used the algorithm to find my answer.

   There are 5.985g of medicine in one dose.

   d. Round the weight of one dose to the nearest gram. Write an equation that shows how to convert the rounded weight to kilograms and solve. Explain your thinking in words.

   5.985g ≈ 6g

   $6g \div 10^3 = 0.006 \text{ kg}$

   $6g = 0.006 \text{ kg}$

   When I divide by 1000 it makes the digits move 3 places to the right because they are getting smaller. Kilograms are 1000 times as large as grams so it takes a lot less to make.